Atomic and Molecular Clusters

There is considerable interest in the study of gas phase and surface-deposited clusters, for many reasons. First, as nanoparticles they constitute intermediates between molecules, with clearly defined energy states (electronic, vibrational, etc.) and condensed matter where these states from continua or bands. It is clearly important to know how bulk properties develop as a function of cluster size. The high ratio of surface atoms to bulk atoms also means that there are many analogies between the chemistry and physics of clusters and of solid surfaces. This has led to studies of the reactivity and catalytic activity of supported clusters. Clusters are also of interest in their own right since, in this intermediate size regime, finite size effects can lead to electronic, magnetic and other properties that are quite different from those of molecules or condensed matter. The use of clusters as components of nanodevices is also attracting much attention.

This book describes the experimental generation, detection and interrogation (spectroscopic and other measurements) of clusters and theoretical approaches developed to aid understanding of their physical properties. Clusters are classified according to their bonding types – ranging from weakly bonded clusters of argon to strongly bonded carbon clusters (the well-known fullerenes) and metal nanoparticles – as well as clusters composed of molecular building blocks. Examples are given of present and possible future applications of clusters in electronic, optical and magnetic devices.

Roy Johnston obtained a first-class honours degree in Chemistry from the University of Oxford (St Catherine's College) in 1983. His doctoral research was carried out in Oxford (1983–6) under the supervision of Professor D. M. P. Mingos. From 1987 to 1989, he was a NATO postdoctoral fellow at Cornell University (with Professor R. Hoffmann) and the University of Arizona (with Professor D. L. Lichtenberger) and, from 1989 to 1995, he was a Royal Society University Research Fellow at the University of Sussex. Appointed as a Lecturer in Inorganic Chemistry at the University of Birmingham in 1995, his research interests include: simulation of the structures, growth and dynamics of clusters; electronic structure calculations on clusters and solids; and application of genetic algorithms to optimization problems in chemical physics. He has published over sixty research papers and reviews in the chemistry and physics literature.

Masters Series in Physics and Astronomy

Series Editor:
David S. Betts
Department of Physics and Astronomy, University of Sussex, Brighton, UK

Core Electrodynamics
Sandra C. Chapman
0-7484-0623-9 (HB)
0-7484-0622-0 (PB)

Atomic and Molecular Clusters
Roy L. Johnston
0-7484-0930-0 (HB)
0-7484-0931-9 (PB)

Forthcoming titles in the series:

Quantum Theory of Solids
Eoin O'Reilly
0-7484-0628-X (HB)
0-7484-0627-1 (PB)

Atomic and Molecular Clusters

Roy L. Johnston

London and New York

Series preface

The *Masters Series in Physics and Astronomy* is aimed squarely at students taking specialized options in topics within the primary areas of Physics and Astronomy, or closely related areas such as Physical Chemistry and Environmental Science. Appropriate applied subjects are also included. The student interest group will typically be studying in the final year of their first degree or in the first year of postgraduate work. Some of the books may also be useful to professional researchers finding their way into new research areas, and all are written with a clear brief to assume that the reader has already acquired a working knowledge of basic core Physics.

The series is designed for use worldwide in the knowledge that wherever Physics is taught at degree level, there are core courses designed for all students in the early years followed by specialized options for those consciously aiming at a more advanced understanding of some topics in preparation for a scientific career. In the UK there is an extra year for the latter category, leading to an MPhys or MSci degree before entry to postgraduate MSc or PhD degrees, whereas in the USA specialization is often found mainly in masters or doctorate programmes. Elsewhere the precise modulations vary but the development from core to specialization is normally part of the overall design.

Authors for the series have usually been able to draw on their own lecture materials and experience of teaching in preparing draft chapters. It is naturally a feature of specialist courses that they are likely to be given by lecturers whose research interests relate to them, so readers can feel that they are gaining from both teaching and research experience.

Each book is self-contained beyond an assumed background to be found in appropriate sections of available core textbooks on Physics and useful Mathematics. There are of course many possibilities, but examples might well include Richard P. Feynman's three-volume classic *Lectures on Physics* (first published by Addison-Wesley in the 1960s) and Mary L. Boas's *Mathematical Methods in the Physical Sciences* (Wiley, 1983). The primary aim of books in this series will be to enhance the student's knowledge base so that they can approach the research literature in their chosen field with confidence. They are not intended as major treatises at the forefront of knowledge, accessible only to a few world experts; instead they are student-oriented, didactic in character and written to build up the confidence of their readers. Most volumes are quite slim and they are generously illustrated.

Different topics may have different styles of questions and answers, but authors are encouraged to include questions at the end of most chapters, with answers at the end of the book. I am inclined to the view that simple numerical answers, though essential, are often too brief to be fully satisfactory, particularly at the level of this series. At the other extreme, model answers of the kind that examination boards normally require of lecturers would make it too easy for a lazy reader to think they had understood without actually trying. So the series style is to include advice about difficult steps in calculations, lines of detailed argument in cases where the author feels that readers may become stuck and any algebraic manipulation which might get in the way of proceeding from good understanding to the required answer. Broadly, what is given is enough help to allow the typical reader to experience the feel-good factor of actually finishing questions, but not so much that all that is needed is passive reading of a model answer.

David S. Betts
University of Sussex

Preface

There is considerable experimental interest in the study of gas phase and surface-deposited metal clusters, for a number of reasons. First, as *nanoparticles* they constitute intermediates between molecules, with clearly defined energy states (electronic, vibrational, etc.) and condensed matter where these states form continua or bands. It is clearly important to know how bulk properties develop as a function of cluster size. The high ratio of surface atoms to bulk atoms also means that there are many points in common between the chemistry and physics of clusters and of solid surfaces. This has led to studies of the reactivity and catalytic activity of supported clusters. Clusters are also of interest in their own right since, in this intermediate size regime, finite size effects can lead to electronic, magnetic and other properties that are quite different from those of molecules or condensed matter. Finally, the use of clusters as components of nanodevices is also attracting much attention.

This book describes the experimental generation, detection and investigation (i.e. spectroscopic and other measurements) of clusters, which are classified according to their bonding types. The importance of theoretical models (especially where experiment cannot unambiguously determine the structures of gas phase clusters) is emphasized. Examples are given of present and possible future applications of clusters in areas such as catalysis and the fabrication of electronic, optical and magnetic nanodevices.

Throughout the book, I have tried to cite the original researchers responsible for the models presented and the results quoted—though this is not a comprehensive review and I apologize for any omissions. At the end of each chapter there is a list of suggested further reading, mainly to review articles wherein readers can find references to the original research literature.

At the end of each chapter, I have included a number of exercises for the interested reader to attempt. Solutions are provided in the appendix.

Finally, in this book I have tried to use the SI units as far as possible—but for cluster dimensions and energies I have tended to adopt the units employed in the original work described here—i.e. ångstroms for length ($1\,\text{Å} = 10^{-10}\,\text{m}$) and electronvolts for energy ($1\,\text{eV} \approx 1.6022 \times 10^{-19}\ \text{J} \approx 96.485\,\text{KJ}\,\text{mol}^{-1}$).

Roy L. Johnston
Birmingham, March 2002

Acknowledgements

I would like to thank the many people who have helped, in various ways, with the preparation, manuscript checking and proof reading of this book and I take responsibility for any errors which remain.

As this book was primarily developed from lecture courses that I have taught over a period of ten years, first at the University of Sussex and latterly at the University of Birmingham, I would like to thank my colleagues with whom I shared the teaching of those courses and who have kindly supplied me with copies of their lecture notes: namely Professor Tony Stace (University of Sussex), Dr Kurt Kolasinski (Queen Mary, University of London) and Dr Mats Andersson (University of Gothenburg). I am grateful to the series editor Dr David Betts (University of Sussex) for going through the manuscript and providing useful advice and feedback and to Dr Grant Soanes and his colleagues at Taylor & Francis for their professional help and encouragement. I would also like to acknowledge past and present members of my research group – in particular Drs Lesley Lloyd, Nicholas Wilson and Christopher Roberts and Mr Thomas Mortimer-Jones – for preparing some of the figures and for helpful comments.

Some of the material in the book was developed from papers and reviews that I have previously written, so I wish to acknowledge all of my co-authors, and in particular Professor Peter Edwards (University of Birmingham). My interest in clusters dates back to my Part II undergraduate project and D.Phil. studies, both carried out in the Inorganic Chemistry Laboratory at the University of Oxford, and I wish to acknowledge my supervisor Professor Mike Mingos for passing on his enthusiasm for clusters and for getting me started in this fascinating area. I have been fortunate to work with and learn from a number of other top scientists and I acknowledge the help and guidance offered to me in particular by Professors Roald Hoffmann (Cornell University), John Murrell (University of Sussex) and Peter Edwards (University of Birmingham). I also acknowledge Professor Richard Palmer (University of Birmingham) for interesting discussions on clusters and nanoscience.

Finally, I thank my wife Suzanne for her love and support and for her patience over while I have been writing this book—which I am pleased to dedicate to her..

Source acknowledgements

I am grateful to a large number of authors and publishers for their permission to reproduce many of the figures that appear in the book (see below). I also wish to thank those authors who have provided me with original hard copy or electronic versions of figures (Professors M. M. Bénard, O. Echt, B. Hartke, P. P. Edwards, K. D. Jordan, R. E. Palmer and S. J. Riley and Drs S. J. Carroll, J. P. K. Doye and D. J. Wales). Many of the sources that I have consulted in the preparation of this book are given in the lists of further reading at the end of each chapter (and references therein). Of all the sources listed, I would like to give special mention to the two-volume work *Clusters of Atoms and Molecules*, edited by Prof. Dr Hellmut Haberland (Springer, Berlin, 1994) and the review article by Prof. Patrick Martin in *Physics Reports* **273**: 199–241 (1996), both of which provided a significant number of the figures reproduced here, as well as being very useful sources of information on experimental and theoretical cluster science.

Figure 1.1 (in part) taken, with permission, from the web page of Prof. Dr B. Hartke http://indy2.theochem.uni-stuttgart.de/~hartke/.

Figures 1.3 and *1.4* from Müller, H. *et al.* (1994) in *Clusters of Atoms and Molecules*, Vol. I (edited by Haberland, H., Berlin: Springer) pp. 114–140 with permission of Springer-Verlag GmbH & Co. KG.

Figures 1.5, 4.1 and *6.14* from Johnston, R. L. (1998) *Philosophical Transactions Royal Society London* **A356**: 211–230 with permission of the Royal Society of London.

Figure 1.7 reproduced with permission from Carroll, S. J. (1999) Ph.D. Thesis, University of Birmingham.

Figure 1.8 from Haberland, H. (1994) in *Clusters of Atoms and Molecules*, Vol. I (edited by Haberland, H., Berlin: Springer) pp. 207–252 with permission of Springer-Verlag GmbH & Co. KG.

Figures 1.9, 4.14(b), 4.14(c), 6.8, 6.9, 7.3 and *7.5* from Martin, T. P. (1996) *Physics Reports* **273**: 199–241 with permission from Elsevier Science.

Figures 2.2, 2.3, 2.6 and *2.8* from Haberland, H. (1994) in *Clusters of Atoms and Molecules* Vol. I (edited by Haberland, H., Berlin: Springer) pp. 374–395 with permission of Springer-Verlag GmbH & Co. KG.

Figure 2.4 from Phillips, J. C. (1986) *Chemical Reviews* **86**: 619–634 with permission. Copyright of the American Chemical Society.

Figure 2.5 from Nelson, D. R. and Spaepen, F. (1989) *Solid State Physics*, **42**: 1–90 with permission of Academic Press.

Figure 2.11 from Lugovoj, E. *et al.* (2000) *J. Chem. Phys.* **112**: 8217–8220. Reprinted with permission. Copyright of the American Institute of Physics.

Figure 3.4 from Nagashima, U. *et al.* (1986) *J. Chem. Phys.* **84**: 207–214. Reprinted with permission. Copyright of The American Institute of Physics.

Figure 3.5 from Wales, D. J. *et al.* (1998) *Chemical Physics Letters* **286**: 65–72 with permission from Elsevier Science.

Figure 3.6 drawn from coordinates produced by Dr D. J. Wales (original citation as for Figure 3.5).

Figure 3.7 taken, with permission, from the web page of Prof. K. D. Jordan http://www.psc.edu/science/Jordan/Jordan.html/ (original citation Pedulla, J. M. *et al.* (1998) *Chemical Physics Letters* **291**: 78–84).

Figure 3.8 from Crofton, M. W. *et al.* (1994) in *Clusters of Atoms and Molecules* Vol. II (edited by Haberland, H., Berlin: Springer) with permission of Springer-Verlag GmbH & Co. KG.

Figure 3.9 from Walker, N. R. *et al.* (1998) *Chemical Physics Letters* **292**: 125–132 with permission from Elsevier Science.

Figure 4.2 from Kappes, M. M. (1988) *Chemical Reviews* **98**: 369–389 with permission. Copyright of the American Chemical Society.

Figures 4.4, 4.5, 4.7, 4.8 and *4.14(a)* from De Heer, W. A. (1993) *Reviews of Modern Physics* **65**: 611–676 with permission. Copyright by the American Physical Society.

Figure 4.9 from Whetten, R. L. & Schriver, K. E. (1989) in *Gas Phase Inorganic Chemistry* (edited by Russell, D. H., New York: Plenum) pp. 193–226 with permission of Kluwer Academic/Plenum Press.

Figure 4.11 from Martin, T. P. *et al.* (1990) *Chemical Physics Letters* **172**: 209–213 with permission from Elsevier Science.

Figure 4.12 from Nishioka, H. *et al.* (1990) *Physical Review B* **42**: 9377–9386 with permission. Copyright by the American Physical Society.

Figure 4.15 from Duff, D. G. *et al.* (1987) *Chemical Communications* pp. 1264–1266 reproduced with permission of the Royal Society of Chemistry.

Figure 4.16 from Cleveland, C. L. *et al.* (1997) *Zeitschrift für Physik* **D40**: 503–508 with permission. Copyright of Springer-Verlag GmbH & Co. KG.

Figures 5.1, 5.2, 5.3 and *5.4* from Riley, S. J. (1994) *Clusters of Atoms and Molecules* Vol. II (edited by Haberland, H., Berlin: Springer-Verlag) with permission of Springer-Verlag GmbH & Co. KG and Argonne National Laboratory, managed and operated by the University of Chicago for the US Department of Energy under contract No. W-31-109-ENG-38.

Figures 5.5, 5.6 and *5.8* from Edwards, P. P. *et al.* (1999) in *Metal Clusters in Chemistry* Vol. 3 (edited by Braunstein, P., Oro, L. A. and Raithby, P. R., Weinheim: Wiley-VCH) pp. 1454–1481 with permission of Wiley-VCH.

Figure 5.7 from Rademann, K. (1989) *Berichte Bunsen-Gesellschaft Physikalische Chemie* **93**: 653–670 with permission.

Figure 5.10 from Collier, C. P. *et al.* (1997) *Science* **277**: 1978–1981 with per-mission. Copyright of the American Association for the Advancement of Science.

Figure 6.1 from Rohlfing, E. A. *et al.* (1984) *Journal of Chemical Physics* **81**: 3322–3330. Reprinted with permission. Copyright of The American Institute of Physics.

Figure 6.2 from Kroto, H. W. (1988) *Science* **242**: 1139–1145 with permission. Copyright of the American Association for the Advancement of Science.

Figures 6.4 and *6.6* from Fowler, P. W. and Manolopoulos, D. E. (1995) *An Atlas of Fullerenes* reproduced with permission. Copyright of P. W. Fowler and D. E. Manolopoulos.

Figure 6.10 from Hargittai, I. (1996) in *Large Clusters of Atoms and Molecules* (edited by Martin, T. P., Kluwer: Dordrecht) pp. 423–435 with permission from Kluwer Academic.

Figure 6.12 from Jarrold, M. F. & Bower, J. E. J. (1992) *Journal of Chemical Physics* **96**: 9180–9190 with permission of the American Institute of Physics.

Figure 6.13 from Rohlfing, C. M. & Raghavachari, K. (1990) *Chemical Physics Letters* **167**: 559–565 with permission from Elsevier Science.

Figure 6.15 from Martin, T. P. & Schaber, H. J. (1983) *Journal of Chemical Physics* **83**: 855–858 with permission of the American Institute of Physics.

Figure 6.16 from Schvartsburg, A. A. & Jarrold, M. F. (2000) *Chemical Physics Letters* **317**: 615–618 with permission from Elsevier Science.

Figure 6.17 reprinted with permission from Hubert, H. *et al.* (1998) *Nature* **391**: 376–378. Copyright of Macmillan Magazines Limited.

Figure 7.2 spectrum from Hudgins, R. R. *et al.* (1997) *Physical Review Letters* **78**: 4213–4216 with permission. Copyright by the American Physical Society. Structure drawings supplied by Dr J. P. K. Doye.

Figure 7.4 from Martin, T. P. (1994) in *Clusters of Atoms and Molecules* Vol. I (edited by Haberland, H., Berlin: Springer) pp. 357–373 with permission of Springer-Verlag GmbH & Co. KG.

Figure 7.6 from Echt, O. (1996) in *Large Clusters of Atoms and Molecules* (edited by Martin, T. P., Kluwer: Dordrecht) pp. 221–239 with permission from Kluwer Academic.

Figure 7.7 from Rohmer, M.-M. *et al.* (1999) in *Metal Clusters in Chemistry* Vol. 3 (edited by Braunstein, P., Oro, L. A. and Raithby, P. R., Weinheim: Wiley-VCH) pp. 1664–1710 with permission of Wiley-VCH.

Figure 8.2 from Schmid, G. (1999) in *Metal Clusters in Chemistry* Vol. 3 (edited by Braunstein, P., Oro, L. A. and Raithby, P. R., Weinheim: Wiley-VCH) pp. 1325–1341 with permission of Wiley-VCH.

Figure 8.3 from Whetten, R. L. *et al.* (1999) *Accounts of Chemical Research* **32**: 397–406 with permission. Copyright of the American Chemical Society.

Figure 8.4 and *8.7* from Simon, U. (1999) in *Metal Clusters in Chemistry* Vol. 3 (edited by Braunstein, P., Oro, L. A. and Raithby, P. R., Weinheim: Wiley-VCH) pp. 1342–1363 with permission of Wiley-VCH.

Figure 8.5 from Sarathy, K. V. *et al.* (1997) *Chemical Communications* 537–538 reproduced by permission of the Royal Society of Chemistry.

Figure 8.8 from Carroll, S. J. *et al.* (2000) *Physical Review Letters* **84**: 2654–2657 with permission. Copyright by the American Physical Society.

Figure 8.9 from Carroll, S. J. *et al.* (1998) *Applied Physics Letters* **72**: 305–307 with permission of the American Institute of Physics.

Figure 8.10 from Tada, T. *et al.* (1998) *Journal of Physics D: Applied Physics* **31**: L21–L24 with permission of IOP Publishing Limited.

Abbreviations

AFM	atomic force microscopy
amu	atomic mass unit
bcc	body-centred cubic
CSE	cluster size effect
DDA	Deposition-Diffusion-Aggregation (model)
DFT	Density Functional Theory
DOS	density of states
EA	electron affinity
EC	Evaporation-Condensation (model)
ESR	electron spin resonance
EXAFS	extended X-ray absorption fine structure spectroscopy
fcc	face-centred cubic
GMR	Giant Magneto-Resistance
hcp	hexagonal close-packed
HOMO	highest occupied molecular orbital
IP	ionization potential (ionization energy)
LB	Langmuir–Blodgett
LDM	Liquid Drop Model
LIF	laser-induced fluorescence
LJ	Lennard–Jones
LMCT	ligand to metal charge transfer
LUMO	lowest unoccupied molecular orbital
MLCT	metal to ligand charge transfer
MO	molecular orbital
MS	mass spectroscopy/mass spectrometry
NMR	nuclear magnetic resonance
PD	Periphery Diffusion (model)
PES	photoelectron spectroscopy
SCA	Spherical Cluster Approximation
SEM	scanning electron microscopy
SET	single electron tunnelling
SHG	second harmonic generation

SIMIT	size-induced metal–insulator transition
SPM	Structureless Packing Model
STM	scanning tunnelling microscopy
STS	scanning tunnelling spectroscopy
TEM	transmission electron microscopy
TOF	time-of-flight
vdW	van der Waals
VRT	vibration-rotation tunnelling
ZEKE	zero kinetic energy (spectroscopy)
ZPE	zero-point energy

CHAPTER ONE

Clusters: Types, Sizes and Experiments

1.1 CLUSTERS

Clusters can be said to bridge the gap between small molecules and bulk materials. In this book, I will describe how clusters are made, what their properties are and how to handle them from a theoretical standpoint. We will examine the relationship of clusters to molecular entities as well as to bulk materials. However, clusters can also be considered to constitute a new type of material, since they often have properties which are fundamentally different from those of discrete molecules or the relevant bulk solid. Though the scientific study of clusters is fairly recent (since the last two decades of the twentieth century), clusters have actually been around for a long time, giving colour to stained glass windows and permitting their beauty to be captured via the medium of photography. In fact, that most symmetrical cluster of all, C_{60}, may be generated in sooting flames and even in space!

1.1.1 What are clusters?

There is still some debate as to what defines a cluster. For the purposes of this book, I will take the term *cluster* to mean an aggregate of a countable number (2–10^n, where n can be as high as 6 or 7) of particles (i.e. atoms or molecules). The constituent particles may be identical, leading to *homo*-atomic (or *homo*-molecular) clusters, A_a, or they can be two or more different species—leading to *hetero*-atomic (*hetero*-molecular) clusters, A_aB_b. These clusters may be studied in the gas phase, in a cluster molecular beam, adsorbed onto a surface or trapped in an inert matrix.

The earliest reference to clusters may have been made by Robert Boyle as long ago as 1661. In his book *The Sceptical Chymist* he spoke of '... minute masses or clusters ... as were not easily dissipable into such particles as compos'd them.'

1.1.2 Types of cluster and cluster bonding

Clusters are formed by most of the elements in the periodic table—even the rare gases! Clusters of the coinage metals (copper, silver and gold) are to be found in stained glass windows and silver clusters are important in photography.

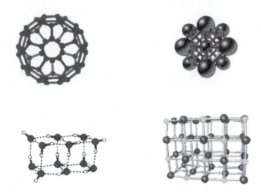

Figure 1.1 Examples of cluster types. Clockwise (from top left): fullerenes; metal clusters; ionic clusters; molecular clusters.

Some clusters (such as water clusters) are even found in the atmosphere! Carbon nanoclusters are now well known, including the famous soccer ball-shaped C_{60} and related fullerenes and the needle-like nanotubes.

Clusters can be subdivided according to the types of atoms of which they are composed and the nature of the bonding in these clusters. Some examples of different types of cluster are shown in Figure 1.1.

1.1.2.1 Metal Clusters

Metallic elements from across the periodic table form a wide variety of clusters. These include: the simple s-block metals (such as the alkali and alkaline earth metals), where the bonding is metallic, delocalized and non-directional, involving primarily the valence *s* orbitals; *sp*-metals (such as aluminium), where the bonding involves both the *s* and the *p* orbitals and has a degree of covalent character; and the transition metals, where the degree of covalency is greater and there is also higher directionality in the bonding, which involves the valence *d* orbitals. Metal clusters may be composed of a single metallic element or of more than one metal, giving rise to the subclass of intermetallic or (nanoalloy) clusters. Theories and models which have been developed to describe metal clusters will be discussed in Chapter 4 and further experimental studies of metal clusters will be treated in Chapter 5.

1.1.2.2 Semiconductor Clusters

Semiconductor clusters are made up of those elements (such as carbon, silicon and germanium) which are semiconductors in the solid state. The bonding in such clusters is covalent and the bonds are strong and directional. This class of cluster also includes compound semiconductor clusters, with polar covalent bonds, such as those formed between gallium and arsenic, Ga_xAs_y. Semiconductor clusters will be discussed in Chapter 6.

1.1.2.3 Ionic Clusters

As the electronegativity difference between two elements, in a compound semiconductor cluster, increases, so the polarity of the bonds increases until a limit is reached where the bonding can better be described as ionic, or electrostatic in nature. In this limit, we have the class of ionic clusters, such as those formed between sodium and chlorine, $[Na_xCl_y]^{(x-y)+}$, magnesium and oxygen $[Mg_xO_y]^{2(x-y)+}$ and so on. Since the anions and cations have fixed charges, the overall charge of the cluster depends on its stoichiometry. Ionic clusters will be discussed in Chapter 7.

1.1.2.4 Rare Gas Clusters

At low temperatures, it is possible to form clusters of the so-called rare (or inert) gas elements (helium to radon). These clusters are bound by weak van der Waals dispersion forces, with the interatomic attraction increasing with increasing atomic mass. In the case of helium, quantum effects, such as superfluidity also manifest themselves. Rare gas clusters will be discussed in Chapter 2.

1.1.2.5 Molecular Clusters

Molecular clusters can be formed by supersonic expansion of molecular vapour. The types of bonding exhibited by molecular clusters include van der Waals bonding, dipole–dipole interactions, higher order multipolar interactions and hydrogen bonding. Molecular clusters will be discussed in Chapter 3.

1.1.2.6 Cluster Molecules

There is a rich chemistry of inorganic and organometallic clusters, which was developed over the second half of the twentieth century. These species, which are generally thermodynamically and/or kinetically stable (with respect to coalescence, etc.) and which can exist in the solid, liquid and vapour phases, include clusters such as P_4, $[B_{12}H_{12}]^{2-}$ and $Os_6(CO)_{18}$. Despite the considerable interest and research activity that they have engendered, these clusters are beyond the scope of this book.

1.1.3 Why Study Clusters?

There is considerable experimental and theoretical interest in the study of elemental clusters in the gas phase and in the solid state. Clusters are of fundamental interest both due to their own intrinsic properties and because of the central position that they occupy between molecular and condensed matter science. One of the most compelling reasons for studying clusters is that they span a wide range of particle sizes, from the molecular (with quantized states), to the microcrystalline (where states are quasi-continuous). Clusters also constitute a new type of material (nanoparticles) which may have properties, which are distinct from those of either discrete molecules or bulk matter.

The study of the evolution of the geometric and electronic structures of clusters and their chemical and physical properties is also of great fundamental interest. One important question, which has arisen, is 'How large must a cluster be before its properties resemble those of the bulk element?' As we shall see, the answer to this question depends critically on which properties are being considered and on the nature of the component atoms. Finally, the high ratio of surface to interior atoms in clusters means that there are many common features between clusters and bulk surfaces.

Some of the interesting features of clusters are discussed briefly below. Most of these will be treated in detail in subsequent sections and chapters.

1.1.3.1 Size Effects

The theoretical and experimental study of the size-dependent evolution of the geometric and electronic structures of clusters, and of their chemical and physical properties, is a major area of current research. The size of a cluster, as well as the type of particles from which it is composed, is an important parameter for tuning the properties of the cluster. Some important, general questions to be addressed are:

- To what extent do cluster properties resemble those of discrete molecules or infinite solids?
- Can the study of large finite clusters tell us anything about the bonding or explain the properties of bulk solids?
- How rapidly do cluster structures and other properties converge towards those of the bulk as the nuclearity (size) increases?
- Can the evolution of band structure with increasing cluster size be detected and, for clusters of metallic elements, at what cluster size is metallic conductivity first observed?
- Can phase transitions be observed and are they of the same type found for bulk solids and surfaces?
- By studying the geometric structures of clusters, how their structures change as a function of size, and cluster growth patterns, can we gain an understanding of crystal growth at the microscopic level?

As well as these general questions, more specific questions can be asked, such as: can a free helium cluster be a superfluid and, if so, what is the smallest number of atoms for which superfluidity can be observed? These questions, and many more, will be addressed in the course of this book.

1.1.3.2 The Cluster–Surface Analogy

Since clusters have a high percentage of their atoms on the surface (see Section 1.2.1), there is a strong link between the chemistry and physics of clusters and of the surfaces of bulk matter. As surface atoms have by definition lower coordination numbers (fewer nearest neighbours) than interior (bulk) atoms, there is the possibility of cluster surface rearrangements, analogous to the reconstructions observed for bulk surfaces, which lower the cluster's surface energy by forming additional surface bonds. Clusters may also be stabilized by the coordination of ligands to their surface.

The reactivity of under-coordinated surface atoms makes clusters of interest as models for heterogeneous catalysis on bulk metal surfaces. In fact, since metal clusters are small metal particles, clusters (generally supported on an inert oxide substrate) can themselves be used as very finely dispersed metal for catalysis.

1.1.3.3 Nanoscience

The field of nanoscience—the study of particles and structures with dimensions of the order of a nanometre (10^{-9} m)—can be said to derive from a visionary article by Richard Feynman in 1960. In this oft-quoted article, entitled '*There's Plenty of Room at the Bottom*', Feynman challenged scientists to develop a new field of study where devices and machines could be constructed from components consisting of a small number (tens or hundreds) of atoms. This article has inspired generations of physicists and chemists to try to make Feynman's vision a reality and there are now a large number of nanoscience laboratories in universities and institutes throughout the world.

A large nanotechnology industry has sprung up in the past decade, motivated by the need to build devices for a variety of electronic, optical, magnetic and even mechanical applications, often using clusters as the basic building blocks. Another rapidly growing area of nanotechnology is the field of organic and bio-organic nanoscience, which encompasses topics such as supramolecular chemistry and molecular recognition, critical for the design of molecular scale machines and computers. This fascinating subject can in part be traced back to the ideas of Eric Drexler and its development has drawn on a number of elegant synthetic and mechanistic studies.

The scientific study of clusters is a relatively new field of research, but it is one that is growing very rapidly. At the beginning of the twenty-first century, clusters promise to play a pivotal role as components in novel electronic, magnetic and optical devices. Physicists, chemists and materials scientists will all contribute to this nanorevolution, as they develop improved methods of synthesis, stabilization and assembly of single nanoclusters and arrays of nanoclusters.

1.1.4 The importance of theory in cluster science

Historically, theory has played (and continues to play) an important role in the development and application of cluster science. Since many cluster properties (e.g. cluster geometries, binding energies and energy barriers) are not easily measured directly from experiment, theoretical models and computational methods have been very useful in helping to interpret spectroscopic (e.g. UV-visible and photoelectron spectroscopy) and mass spectrometric data.

The field of clusters also serves as an exacting testing ground for theoretical methods—testing the range of validity of theoretical models derived from the extremes of atomic/molecular and solid state physics. One of the challenges for theory is to come up with a theory of cluster structure and bonding which is applicable over an extremely large size range—from a few atoms to millions of atoms.

1.2 VARIATION OF CLUSTER PROPERTIES WITH SIZE

One of the fascinating aspects of clusters concerns the evolution of cluster properties as a function of size. This section introduces approximations and models which have been introduced to describe this evolution.

1.2.1 The Spherical Cluster Approximation

In the Spherical Cluster Approximation (SCA), an N-atom cluster is modelled by a sphere. This becomes a better approximation as the cluster gets larger and enables equations to be derived for the number of surface atoms and the fraction of surface atoms, in the large cluster limit.

For a spherical cluster composed of N atoms, the cluster radius (R_c), surface area (S_c), and volume (V_c) can be related to the radius (R_a), surface area (S_a) and volume (V_a) of the constituent atoms, as follows.

First the cluster volume is assumed to be simply the volume of an atom multiplied by the number of atoms in the cluster:

$$V_c = NV_a \tag{1.1}$$

Of course, this represents a significant oversimplification, since it does not take into account the fact that hard spheres cannot pack to fill space exactly, but as we are interested in deriving scaling relationships, we shall not worry about the packing fraction here (but see Section 4.6.2). Expressing Equation (1.1) in terms of the cluster and atomic radii, we have:

$$\tfrac{4}{3}\pi R_c^3 = N \tfrac{4}{3}\pi R_a^3 \tag{1.2}$$

which can be rearranged to give the following relationship between cluster and atomic radius:

$$R_c = N^{1/3} R_a \tag{1.3}$$

The surface area of the cluster is then related to that of an atom as follows:

$$S_c = 4\pi R_c^2 = 4\pi \left(N^{1/3} R_a\right)^2 = N^{2/3} S_a \tag{1.4}$$

In the limit of a large cluster, the number of surface atoms (N_s) in a cluster is given by dividing the surface area of the cluster by the cross-sectional area of an atom (A_a):

$$N_s = \frac{4\pi N^{2/3} R_a^2}{\pi R_a^2} = 4N^{2/3} \tag{1.5}$$

As will become evident below (and in later chapters), many properties of clusters depend on the fraction of atoms (F_s) which lie on the surface of the cluster. For pseudo-spherical clusters this quantity is given by:

$$F_s = \frac{N_s}{N} = 4N^{-1/3} \tag{1.6}$$

Figure 1.2 shows the fraction of surface atoms (F_s) plotted against $N^{-1/3}$ for clusters consisting of concentric icosahedral shells of atoms. Such geometric shell

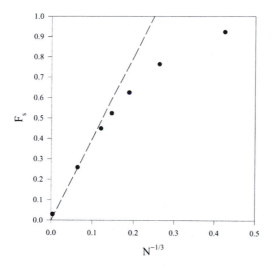

Figure 1.2 Fraction of surface atoms (F_s) plotted against $N^{-1/3}$ for icosahedral geometric shell clusters. The dashed line represents the prediction of the Spherical Cluster Approximation.

clusters are commonly observed for rare gas (Chapter 2) and metal (Chapter 4) clusters. The figure shows that the fraction of surface atoms approaches the limiting value ($4N^{-1/3}$) given by Equation (1.6) as the cluster size increases.

1.2.2 Cluster Size Regimes

In this book, I shall refer to three cluster size regimes: small clusters (< 100 atoms); medium-sized clusters (100–10,000 atoms); and large clusters ($> 10,000$ atoms).

From the Spherical Cluster Approximation (Equation (1.3)), the diameter (D) of a cluster of N atoms is given by:

$$D = 2R_c = 2R_{WS}N^{1/3} \tag{1.7}$$

where R_{WS} is the Wigner–Seitz radius of the element under consideration (i.e. the radius of a sphere whose volume is equal to the volume per atom in the solid). Table 1.1 gives the ranges of diameters for the three cluster size regimes, taking as an example clusters of sodium atoms (with $R_{WS} = 0.2$ nm)

Considering the fraction of atoms on the surface of the cluster, ranges of F_s values, for the three cluster size regimes are also given in Table 1.1, from which it is apparent that clusters with as many as 10,000 atoms still have nearly 20% of their atoms on the surface. In fact, F_s only drops below 0.01 (i.e. less than 1% of atoms are on the surface) for $N > 6.4 \times 10^7$ atoms, which corresponds to a diameter of approximately 0.16 μm for sodium clusters.

Table 1.1 Classification of clusters according to size. N is the number of atoms, D is the diameter (for a cluster of sodium atoms) and F_s is the fraction of surface atoms.

	Small	Medium	Large
N	$\leq 10^2$	10^2–10^4	$> 10^4$
D / nm	≤ 1.9	1.9–8.6	> 8.6
F_s	≥ 0.86	0.86–0.19	≤ 0.19

1.2.3 Cluster Size Effects

1.2.3.1 Size-Dependent Melting

As long ago as 1871, Lord Kelvin posed the question: 'Does the melting temperature of a small particle depend on its size?' Nearly 40 years later, Pawlow predicted that the melting temperature of a metallic particle should decrease as the particle gets smaller, though the unambiguous verification of the prediction had to wait until 1976, when Buffat and Borel measured the melting temperatures of gold clusters under a transmission electron microscope. Melting temperatures for the smallest nanoclusters turned out to be as low as 300 K—significantly lower than the melting point (approximately 1336 K) of bulk elemental gold. (See Figure 1.4 in Section 1.2.4 for a plot of melting temperature vs. cluster radius for gold clusters.) The subject of cluster melting is a complex one. There is evidence that cluster melting may not be a sharp (first order) phase transition (as for bulk metals), but rather may be a continuous (second order) phase transition. Berry has shown, theoretically, that in clusters at around the melting point there can be coexistence of liquid-like (surface) regions and solid-like (bulk) regions. Electron microscopy studies of surface-deposited 50 nm diameter Sn and Pb clusters have verified these predictions, showing a solid core surrounded by a liquid outer layer.

1.2.3.2 General Considerations

Lord Kelvin's specific enquiry can be generalized as the following problem: 'How large must a cluster of atoms be before its properties resemble those of the bulk element?' The answer to this apprently simple question is exceedingly complex, and depends on the type of atoms constituting the cluster and the type of physical property that is being measured. The variation of cluster properties with size can be gathered under the heading Cluster Size Effects (CSEs).

Many properties of clusters depend on the fraction of atoms lying on the surface of the nanocluster. As mentioned above, the percentage of surface atoms only drops below 1% for clusters of over 64 million atoms. It is this high surface to bulk ratio which has made metal nanoclusters so important, for example, as a source of finely dispersed metal for application in heterogeneous catalysis. More generally, the size-dependent behaviour of cluster properties suggests the very exciting prospect of using nanoclusters as building blocks to construct electronic, magnetic or optical devices with characteristics that can be fine-tuned by carefully controlling the size of the component nanoclusters.

1.2.4 Scaling Laws

In the large cluster regime, many generic cluster properties (G), such as ionization energy (ionization potential, IP), electron affinity (EA), melting temperature (T_m) and cohesive or binding energy (E_b) show a regular variation with cluster size. By adopting a spherical cluster model, in which the N-atom cluster is approximated by a sphere of radius R, this smooth CSE behaviour can be described by simple scaling laws (interpolation formulae), either in powers of the cluster radius:

$$G(R) = G(\infty) + aR^{-\alpha} \qquad (1.8)$$

or the nuclearity (number of atoms):

$$G(N) = G(\infty) + bN^{-\beta} \qquad (1.9)$$

where $G(\infty)$ is the value of property G in the bulk limit. Since many properties depend on the ratio of surface to bulk (volume) atoms in a cluster, and since $F_s \propto N^{-1/3}$ ($\propto 1/R$), the exponents in Equations (1.8) and (1.9) are generally $\alpha = 1$ and $\beta = \frac{1}{3}$.

As an example of the application of scaling laws, the ionization energies of potassium clusters, with up to 100 atoms, can be fitted to a high degree of accuracy by the following interpolation formulae:

$$IP^K(N)/eV = 2.3 + 2.04N^{-1/3} \qquad (1.10)$$

$$IP^K(R)/eV = 2.3 + 5.35\,(R/\text{Å})^{-1} \qquad (1.11)$$

A plot of IP^K against N for potassium clusters, including a fit to Equation (1.10), is shown in Figure 1.3. The variation of ionization energies and electron affinities of metal clusters will be detailed in Chapter 4, in the context of the Liquid Drop Model.

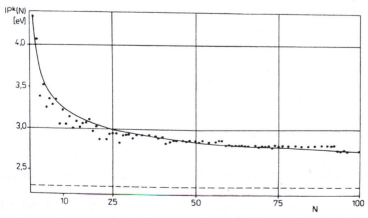

Figure 1.3 The The ionization energy of potassium clusters as a function of cluster size N. The points represent experimental data, the solid line is taken from the interpolation formula (Equation (1.10)) and the dashed line indicates the bulk work function.

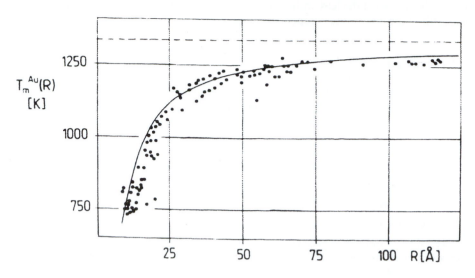

Figure 1.4 The melting temperature of gold clusters as a function of radius R. The points represent experimental data, the solid line is taken from the interpolation formula (Equation (1.12)) and the dashed line indicates the bulk melting temperature.

As discussed above, the melting temperatures (T_m^{Au}) of gold clusters have been found to decrease with decreasing cluster size, following a $1/R$ trend, and showing a good fit to the expression:

$$T_m^{Au}(R)/K = 1336.15 - 5543.65\,(R/\text{Å})^{-1} \tag{1.12}$$

as shown in Figure 1.4.

As can be seen from Figures 1.3 and 1.4, while large clusters show good agreement with scaling laws, for smaller clusters some deviations are observed. In fact, large deviations (usually oscillations about the smooth CSE trend representing the power law) are observed for many properties in the medium- and (especially) the small-cluster size regimes. Such deviations arise due to Quantum Size Effects (QSEs, such as those caused by electronic shell closings) and Surface Effects (geometric shell closings), which will be described in Chapter 4. A schematic representation of a cluster property (G), which exhibits smooth CSE (or Liquid Drop) behaviour at high nuclearity and Quantum Size and Surface Effect oscillations at low nuclearity, is shown in Figure 1.5.

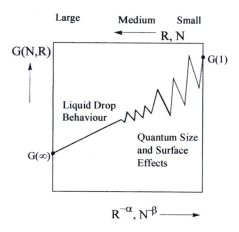

Figure 1.5 Schematic representation of the variation of a generic cluster property (G) with cluster size.

1.3 CLUSTER EXPERIMENTS

Cluster experiments can be divided into three main stages: *cluster generation* (production of clusters); *cluster investigation* (where the clusters are probed using spectroscopic or other techniques); and *cluster detection* (generally using mass spectrometry). This section presents examples of specific experimental techniques for generating and studying different types of clusters, though further examples of cluster experiments will be presented in later chapters. The treatment given here will not be exhaustive and readers are directed to the references at the end of this chapter for more detailed discussions of cluster experiments and for diagrams of experimental apparatus.

It is important to note that progress in cluster science has followed closely on the development of new experimental techniques. In particular, the development of molecular beam techniques has enabled the study of 'free' clusters in an interaction-free environment. The study of free clusters, however, presents a number of problems associated with difficulties in measuring physical properties of single particles and with generating intense size-selected cluster beams. Also, since there are often many cluster isomers with similar energies and low barriers to interconversion, the concept of 'cluster structure' itself may not always be well defined. Another class of experiments involves the deposition of size-selected clusters on a substrate (such as graphite, silicon or an inorganic oxide), as shown schematically in Figure 1.6, or in an inert gas matrix. These experiments will be discussed further in Chapter 8. While such experiments allow individual clusters to be studied by microscopic techniques, it is difficult to infer the geometric or electronic structure of a free cluster from that of the corresponding surface-supported cluster, since such clusters may be perturbed by the substrate. The same caveat applies to the study of thin films and crystals of ligand-passivated clusters (see Chapter 8).

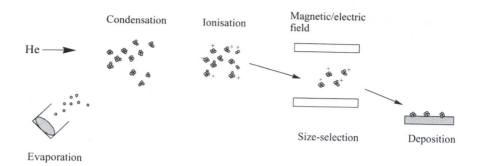

Figure 1.6 Schematic representation of cluster generation (by condensation in flowing helium), size selection and deposition.

Finally, since clusters in a molecular beam are generally not in thermodynamic equilibrium, the concept of temperature is not well defined for clusters and many researchers prefer to avoid applying terminology which has thermodynamic connotations (i.e. which imply well defined phases in equilibrium). Berry has shown, however, that clusters can exist in a number of phase-like forms, some of which (such as the liquid-like or solid-like forms) have similar properties to the bulk phases, while others are unique for clusters—having no bulk counterpart.

1.3.1 Cluster Sources

The first stage in a cluster experiment is the generation of the clusters. Clusters are generated in a *cluster source*, a number of which are described below. In all cluster sources, cluster generation consists of the processes of: *vaporization* (production of atoms or molecules in the gas phase); *nucleation* (initial condensation of atoms or molecules to form a cluster nucleus); *growth* (the addition of more atoms or molecules to the initially formed nucleus); and *coalescence* (the merging of small clusters to form larger clusters). As well as growing, clusters can also shrink by evaporation or fragmentation. Depending on the nature and conditions of the source, different size distributions of clusters may be generated, as described below.

1.3.1.1 Knudsen Cell (Effusive Source)

The Knudsen cell, or effusive source, is a continuous, low flux, subsonic source which involves heating a volatile solid or liquid in an oven with a small aperture, thereby generating a low vapour pressure. At low vapour pressures the mean free path of particles in the cell is greater than the diameter of the aperture, so there are very few collisions before leaving the cell. The resolution of effusive sources is poor, as broad energy and angular distributions are obtained.

If the aperture is small enough (though nozzle clogging may be a problem for less volatile solids) that the solid (or liquid)–gas equilibrium is not perturbed, an equilibrium distribution of clusters is generated, with a Maxwell–Boltzmann distribution of energies. The overall trend is that the cluster intensity (I) falls exponentially with increasing cluster nuclearity (N):

$$I(N) = ae^{-bN} \tag{1.13}$$

so that small clusters predominate. However, abundance spectra have structure superimposed on this trend, with intensities related to the binding energies of the clusters (see Section 1.3.5.2). In the case of antimony, for example, a larger mass spectral peak is always observed for the Sb_4 cluster than for either Sb_3 or Sb_5.

1.3.1.2 Supersonic (Free Jet) Nozzle Sources

Supersonic nozzle sources (also known as free jet condensation sources) are of two main types. Sources for generating clusters of inert gases, molecules and low-boiling metals (such as mercury) generally operate without a carrier gas. The seeded supersonic nozzle source, on the other hand, is used to produce intense cluster beams of relatively low-melting metals, such as the alkali metals. The metal is vaporized (with a vapour pressure of 10–100 mbar $\approx 10^3$–10^4 Pa) in an oven and the vapour is seeded in an inert carrier gas at a stagnation pressure (p_0) of several atmospheres ($\approx 10^5$–10^6 Pa) and a temperature (T_0) of 77–1500 K. The metal/carrier gas mixture is then expanded through a small nozzle (with a diameter of 0.03–1 mm) into a vacuum (10^{-1}–10^{-3} Pa), thereby creating a supersonic beam. Slit nozzles have also been used to generate two-dimensional cluster beams. Such beams are preferred for optical measurements, where longer path lengths are required.

The high backing or stagnation pressure in the condensation region and the small diameter of the nozzle means that the particle mean free path is much smaller than the nozzle diameter. Thus, there are many collisions during the expansion phase. The expansion, which is adiabatic and isenthalpic, cools the vapour, which becomes supersaturated, and clusters condense.

Cluster growth typically ceases beyond a few nozzle diameters from the nozzle exit, when the vapour density becomes too low. The mean velocities of the clusters (and hence their kinetic energies) increase but random thermal motions are reduced. This means that the velocities of the clusters, relative to each other, are very low—corresponding to low beam temperatures. Cluster cooling continues until cluster flow becomes molecular, rather than hydrodynamic. Further cooling can occur by evaporation of atoms from the cluster.

These sources produce intense, continuous cluster beams with narrow velocity distributions, and clusters with hundreds or thousands of atoms can be generated. Because evaporative cooling takes place, very low temperatures (near the evaporation limit) can be generated and cluster abundances are sensitive to cluster binding energies and hence are highly structured, being related to electronic or packing properties of the clusters. Three factors govern the cluster content and size of clusters from free jet sources: the stagnation pressure (p_0); the temperature (T_0) and the nozzle aperture cross-section (A). The cluster content (number of clusters) and average cluster size $\langle N \rangle$ increase with increasing p_0 and A, while they decrease with increasing T_0.

In the absence of a carrier gas, small clusters (with fewer then ten atoms) generally form. Temperatures exceeding 1,500 K have been attained, with the higher temperatures being required to generate lithium clusters—though this method is not useful for more refractive elements. For mercury and alkali metal clusters, temperatures in the range 600–1,300 K are generally used, while for rare gas clusters (due to the weaker binding of rare gas clusters) temperatures well below room temperature are normally required for cluster condensation.

1.3.1.3 Laser Vaporization–Flow Condensation Source

The laser vaporization–flow condensation source (which was introduced by Smalley and co-workers in the early 1980s) is a pulsed cluster source which is used to produce small- and medium-sized (up to several hundreds of atoms) clusters of any metal—including refractory transition metals, carbon and silicon. It combines laser ablation with a supersonic jet expansion. Though the time averaged flux is smaller than in the seeded supersonic nozzle source, cluster intensities within each pulse are high. In pulsed cluster sources, skimmers, which seperate regions of differing pressure also serve to prevent the formation of shock waves.

In the laser vaporization source, vapour is produced by pulsed-laser ablation of a rod of the material of interest, using an intense ($> 10^7$ W cm^{-2} or 10 mJ mm^{-2}) pulsed UV Nd:YAG or excimer laser. Each 10 ns pulse vaporizes 10^{14}–10^{15} atoms per mm^2 of the target. The vaporized material (a plasma with a temperature of around 10^4 K) is introduced into a pulse of cold helium, which rapidly cools the vapour and induces cluster formation. The adiabatic cooling occurs at a greater rate than for the seeded supersonic nozzle and temperatures of under 100 K can be produced. Further evaporative cooling can then occur. The use of a laser for cluster generation also leads to some cluster ionization, so this source generates neutral, cationic and anionic clusters.

1.3.1.4 Pulsed-Arc Cluster Ion Source (PACIS)

PACIS sources are closely related to laser vaporization sources, but the cluster material is instead vaporized by an intense electrical discharge. Cluster beams generated in this way are significantly more intense than when using laser vaporization, with deposition rates of 0.2 nm per pulse having been reported for the deposition of lead clusters. As many as 10% of the clusters generated by this source are charged.

1.3.1.5 Ion Sputtering Source

In an ion sputtering source, clusters are produced by bombarding a target with high energy inert gas ions, as shown in Figure 1.7a. Clusters of refractory metals and other high-melting materials can be generated. This source typically generates small, singly ionized clusters, although mass selected beams of clusters with hundreds of atoms can be generated. Clusters are generated hot and cluster cooling involves evaporation, so that the abundance spectra reflect the thermodynamic stabilities of the clusters, via their binding energies.

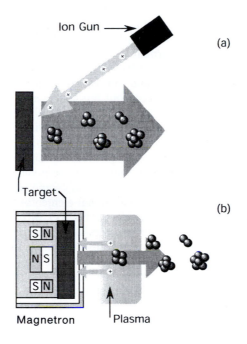

Figure 1.7 Sputtering cluster sources: (a) ion sputtering—using an ion gun to accelerate ions onto the target; (b) magnetron sputtering—using a magnetically confined plasma as an ion source.

The heavier inert gases (Kr and Xe) are generally used as sputtering ions, with bombardment energies in the range 10–30 keV and currents of approximately 10 mA.

1.3.1.6 Magnetron Sputtering Source

The magnetron sputtering source (Figure 1.7b) is a modified type of sputtering source, wherein a plasma is ignited in argon over a target by applying a d.c. or r.f. potential and is confined using a magnetic field. Argon ions are then accelerated onto the target, resulting in sputtering, as described above. Magnetron sputtering sources generally produce small clusters (2–30 atoms), with cluster intensity falling off rapidly with increasing size. Palmer and co-workers, however, have recently developed a hybrid magnetron sputtering/condensation source, which is capable of producing a high flux of Cu and Ag, clusters with sizes ranging from two to as many as 70,000 atoms.

1.3.1.7 Gas-Aggregation/Smoke Source

In the gas-aggregation or smoke source, metal vapour is generated by evaporation or sputtering and is introduced into a cold inert quench gas (He or Ar at a pressure of 50–500 Pa), where the vapour becomes supersaturated and clusters aggregate. Stagnation and aggregation can occur in the same chamber or gentle expansion can occur, through an aperture of 1–5 mm diameter, into an aggregation chamber with several hundred Pa pressure of cold quench gas. The mechanism of cluster growth and aggregation in the gas aggregation source is analogous to cloud and smoke formation.

The low temperature of the inert gas ensures that cluster growth consists primarily of single-atom addition and that evaporation is negligible. Because of this, the observed cluster abundances from this source are determined by collision statistics and are featureless, smooth functions of cluster size. Final cluster temperatures are generally lower than those from heavily clustered seeded beams, though the clusters in lightly clustered, seeded supersonic beams may be cooler.

Gas-aggregation cluster sources produce continuous beams of low-to-medium boiling (< 2,000 K) metals, including the noble metals. Cluster beam intensities are much lower than for seeded supersonic nozzle sources, but very large (> 20,000 atoms) clusters (of alkali metal elements) can be generated using gas-aggregation sources, due to the greater number of collisions with the high density of the cooling gas. Smaller clusters can be generated by streaming the quenching gas past the vaporization zone and controlling the pumping rates. The cluster size distribution also depends on the physical dimensions of the cell, as this determines the time during which cluster growth can occur. A subsonic beam leads to a low intensity, broad velocity distribution, which is internally hot.

Figure 1.8 shows a smoke source, used for the production of C_{60} and other fullerene clusters (see Chapter 6).

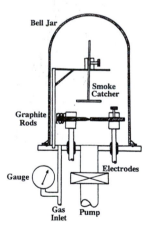

Figure 1.8 A smoke source used for the production of C_{60} and other fullerene clusters.

1.3.1.8 Liquid Metal Ion Sources

Liquid metal ion sources are used to produce multiply charged clusters of low melting metals. In field desorption sources, very high electric fields are generated at the tip of a needle which has been wetted in the metal of interest and heated above the melting point of the metal. The initially hot, multiply charged clusters undergo evaporative cooling and cluster fission, to generate smaller clusters.

1.3.1.9 Spray Sources

Spray sources are used for generating clusters from liquids and solutions. The *electrospray source* generates solvated ion clusters by the injection of a solution through a needle (which can carry a relative positive or negative potential) into a stagnation chamber with a flowing inert carrier gas. This method allows complex, involatile molecules to be obtained in the gas phase as solvated ions.

The *thermospray source* is also used to generate cluster beams of involatile materials from liquids and solutions. The liquid sample is partially pyrolyzed before expansion, generating a subsonic beam containing neutral and ionic clusters.

1.3.2 Cluster Formation

In cluster experiments, the extent of clustering depends on many factors—for example, in supersonic beam experiments, important factors are: the stagnation pressure; the diameter of the nozzle; and the carrier gas temperature—with lower temperatures and higher pressures leading to larger clusters. In this section, I will briefly discuss the important stages of cluster nucleation and growth, as well as mechanisms by which clusters can cool down.

1.3.2.1 Cluster nucleation

If the local thermal energy of the beam is less than the binding energy of the dimer, then a three-atom collision can lead to the formation of a dimeric nucleus, with the third atom removing the excess energy as kinetic energy:

$$A + A + A\ (KE_1) \rightarrow A_2 + A\ (KE_2 > KE_1)$$

In the presence of an excess of cold, inert carrier or quench gas (B), the nucleation step is more efficient:

$$A + A + B\ (KE_1) \rightarrow A_2 + B\ (KE_2 > KE_1)$$

The dimer then acts a site for further condensation (cluster growth). By increasing both the nozzle diameter and the stagnation pressure, the average cluster size and density increase, as the beam character changes from Knudsen-like to free jet-like.

1.3.2.2 Cluster Growth

The initially formed cluster nucleus acts as a seed for further cluster growth. Early growth occurs by accretion of atoms (or molecules) one at a time. Subsequently, collisions between smaller clusters can lead to coalescence and the formation of larger clusters:

$$A_N + A \rightarrow A_{N+1}$$
$$A_N + A_M \rightarrow A_{N+M}$$

In the cluster growth region, the clusters are generally quite hot, so there is competition between growth and decay—i.e. clusters shrinking by losing individual atoms (evaporation) and/or fragmentation (splitting into two or more clusters). These processes are merely the reverse of those shown above.

1.3.2.3 Cluster Temperature

It is difficult to measure and define cluster temperature accurately. If there is negligible clustering then cluster temperatures are very low. Cluster growth, which is an exothermic process, causes an increase in the temperature (i.e. the internal energy increases due to the heat of condensation of the added atoms). In heavily clustered beams, where there is a high ratio of atoms to clusters (and the clusters are larger), the clusters are very hot (possibly molten) when generated.

1.3.2.4 Cluster Cooling

There are three main mechanisms by which clusters, in a molecular beam can lower their temperature (i.e. lower their internal energy).

(a) Collisional cooling. Collisions with other atoms in the beam remove the excess energy as kinetic energy:

$$A_N (T_1) + B (KE_1) \rightarrow A_N (T_2 < T_1) + B (KE_2 > KE_1)$$

where B may be another atom of element A or an inert carrier gas atom. Using a cold carrier gas leads to more efficient collision cooling. This cooling mechanism is only significant in the condensation and initial expansion regions.

(b) Evaporative cooling. Clusters can lower their internal energies by evaporation—i.e. by losing one or more atoms in an endothermic desorption process. In order for evaporation to occur, internal energy must be chanelled into the appropriate cluster vibrational mode(s), in order to overcome the kinetic activation barrier to bond breaking. After evaporation, this excess energy is imparted as kinetic energy to the escaping atom

$$A_N (T_1) \rightarrow A_{N-1} (T_2 < T_1) + A (KE) \rightarrow A_{N-2} (T_3 < T_2) + A (KE) \rightarrow \dots$$

This mechanism, which forms the basis of the Klots evaporative ensemble, is the only cooling mechanism once free flight has been achieved (i.e. when collisions no longer occur).

(c) Radiative cooling. Finally, clusters can lower their internal energies by emitting infrared radiation:

$$A_N (T_1) \rightarrow A_N (T_2 < T_1) + h\nu$$

Radiative cooling, however, is an inefficient cooling mechanism, which is slow compared with the time scales of typical cluster experiments.

1.3.2.5 Distribution of Cluster Sizes

The distribution of cluster sizes produced by a cluster source depends on a number of factors. First, as discussed above, the distribution is strongly influenced by the particular cluster source being used. The size distribution depends on the way in which the vapour is generated, the initial temperature and pressure, the presence (and backing/stagnation pressure and temperature) of a carrier or quenching gas, the dimensions and shape of apertures and nozzles and whether or not supersonic expansion is performed. As will be discussed in Section 1.3.5.2, these experimental conditions can strongly influence the observed abundances in the mass spectra measured for a beam of clusters.

1.3.3 Generation of Cluster Ions

Many experiments involving clusters rely on being able to separate them according to their mass (i.e. how many atoms they possess). In order to do this, it is generally necessary to ionize the clusters so that mass selection (see below) can be accomplished by deflecting the clusters in a magnetic or electric field. Depending on the material and the type of experiment, either cations or anions may be created. As described above, cations and/or anions may also be formed in the initial cluster generation step—for example when using laser ablation, electric discharge or sputtering sources.

1.3.3.1 Generation of Cluster Cations

Positively charged clusters can be generated from neutral clusters, either by electron impact or by photoionization of the corresponding neutral clusters.

(a) Electron impact ionization, using either thermal electrons (generated by thermionic emission from a heated wire) or a focussed electron beam, creates cluster cations by knocking electrons off the neutral clusters. It should be noted that electron impact ionization is often accompanied by cluster fragmentation because of the large geometry changes that accompany ionization and because of excess energy arising from the ionization process:

$$A_N + e^- (KE) \rightarrow (A_N^+)^{**} + 2e^- \rightarrow (A_{N-M}^+)^* + A_M + 2e^- \rightarrow \ldots$$

and further fragmentation may subsequently occur. The fact that electron impact ionization can lead to fragmentation should be remembered when interpreting mass spectroscopic abundances of cationic clusters.

(b) Photoionization (using a laser light source) is usually a gentler way of generating cationic clusters. It has the advantage that photon frequencies can be tuned across quite a wide range. Low energy visible or near UV photons can cause ionization without fragmentation, as there is not sufficient energy to break cluster bonds. (Running the ionizing laser at low power reduces the probability of multi-photon absorption, which can also lead to cluster fragmentation.) A variant on this type of experiment is to use near-threshold frequency light, where the laser frequency is tuned until the photons have just enough energy to ionize the

clusters. In this way, the appearance potential of a cluster cation (of a given mass), and hence the ionization energy of the corresponding neutral cluster, can be determined.

(c) Electric discharge—positively charged clusters can be generated in free jet sources, by using a corona discharge within the stagnation chamber, prior to expansion.

1.3.3.2 Generation of Cluster Anions

There are two main ways of generating anionic clusters, by electron transfer or by capture of low energy electrons.

(a) Electron transfer is achieved by having electropositive alkali metal atoms (M) (such as rubidium or caesium) in the vapour phase. The transfer of the valence electron from the alkali metal to the cluster:

$$A_N + M \rightarrow A_N^- + M^+$$

is thermodynamically favoured if the ionization energy of the cluster is greater than that of the alkali metal atom: $IP(A_N) > IP(M)$. This type of process is favoured for creating cluster anions of metals such as Cu, Ag and Au, where the clusters have relatively high IPs. One drawback of the method is the possible inclusion of alkali metal atoms into the cluster.

(b) Capture of low energy electrons can also be used to generate anionic clusters:

$$A_N + e^- \rightarrow A_N^-$$

This method can be used for a wider range of clusters than the electron transfer method and does not risk incorporation of unwanted atoms into the cluster.

1.3.4 Mass Selection and Detection of Clusters

Cluster ions (both positively and negatively charged) are easily separated and selected, usually according to their masses—i.e. by mass spectrometry. A number of mass spectrometers, differing in the nature of the mass analyser, are currently in use.

1.3.4.1 Wien Filter

In the Wien filter, mass separation is performed by crossed homogeneous electric (E) and magnetic (B) fields, which are both perpendicular to the direction of the ionized cluster beam. The net force on a cluster with charge Q, mass M and velocity v ($\propto M^{-1/2}$) vanishes if $E = Bv$. The cluster ions are accelerated by a voltage V to an energy QV. Clusters with a mass to charge ratio (M/Q) of $2V/(E/B)^2$ are undeflected by the Wien filter and it is these undeflected ions that are selected (by collimators) and detected.

The resolution of the Wien filter depends on the spread of velocities of the cluster ions (which depends on the initial velocity distribution), the strength of the electric and magnetic fields, and the collimator widths. Higher resolution is obtained by using high field strengths and narrow collimators. Increasing the acceleration voltage also reduces the effect of the initial velocity distribution. The mass resolution ($\delta M/M$) of commercial Wien filters is typically of the order of 10^{-2} (i.e. 1% of the mass), with a mass range of 1–5,000 amu, though the resolution is worse for the larger clusters. The mass range is mainly limited by field strengths.

1.3.4.2 Magnetic Sector Mass Spectrometer

Traditional mass spectrometers use homogeneous electric or magnetic field sectors to deflect charged clusters by an extent depending on their charge-to-mass ratio and their velocities. In some groups, double sector instruments have been developed, in which cluster cations are accelerated to the same kinetic energy and deflected by a magnetic sector field so that only one mass, with the correct momentum can exit the magnetic sector. The cations are then passed through an electric sector field, with an energy dispersion adjusted to compensate for that of the magnetic sector. This results in double focussing of the mass-selected cluster beam, high resolution and very good background suppression.

1.3.4.3 Quadrupole Mass Filter

In the quadrupole mass filter, an a.c. electric field is superimposed on a d.c. electric field, with the quadrupole formed by four cylindrical metal rods held in a square geometry, with bias voltages $V = \pm (V_{dc} + V_{ac} \sin(\omega t))$, such that adjacent rods have equal and opposite voltages. By carefully varying the field strengths (V_{dc} and V_{ac}) and the a.c. frequency (ω), a stable helical trajectory through the filter is possible for cluster ions within a narrow mass range. Mass resolution is affected by the field strength, the cluster velocity and the alignment of the rods. Resolutions ($\delta M/M$) of 10^{-3} (i.e. an order of magnitude better than the Wien filter) are typical. Masses of up to 9,000 amu can be detected by using high voltages and lower a.c. frequencies, though increase in the mass range is generally achieved at the expense of mass resolution.

A variation of the quadrupolar mass filter is the so-called *phase space compressor*, in which the quadrupole region is filled with an inert buffer gas such as helium or argon. The cluster ions lose internal energy (and are therefore cooled) in collisions with the buffer gas, which also serves to damp out differences in the initial cluster kinetic energies so that only the translational energy due to the fields within the quadrupole remain.

1.3.4.4 Time-of-Flight Mass Spectrometry

In the *time-of-flight* (TOF) mass spectrometer, cluster ions are accelerated by a succesion of homogeneous electric fields (an ion gun) into a field-free flight tube, finally impacting on an ion detector. The mass-to-charge ratio (M/Q) of the cluster is determined from the measured time-of-flight. Mass resolution is limited by initial conditions (velocities etc.), timing accuracy and power supply stability, but the resolution of the analyser is not mass-dependent and the resolution ($\delta M/M \approx 10^{-4}-10^{-3}$) is good throughout the mass range. One disadvantage of the TOF mass spectrometer is that it only operates on short cluster pulses and thus low ion intensities are detected.

The mass resolution of the TOF mass spectrometer can be improved by reducing the effect of the initial conditions, as in the reflectron scheme, wherein the cluster ions are reflected back along the flight tube using an electrostatic mirror. Since fast ions take longer to reverse their direction than slower ions, the time spread of clusters of a given mass is reduced by this process and all ions (of the same mass) arrive at the detector almost simultaneously. Mass resolutions ($\delta M/M$) of 10^{-5} have been achieved with reflectron TOF devices (approximately an order of magnitude better than is possible in a normal TOF spectrometer).

Figure 1.9 shows the apparatus used by Martin and colleagues for the production, photoionization and TOF mass analysis of a variety of clusters.

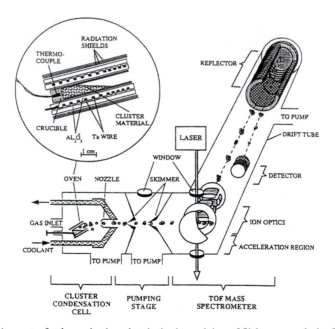

Figure 1.9 Apparatus for the production, photoionization and time-of-flight mass analysis of clusters.

1.3.4.5 Ion Cyclotron Resonance Mass Spectroscopy

Ion cyclotron resonance mass spectroscopy is performed by confining cluster ions in a three-dimensional ion trap, using a combination of a static quadrupolar electric field and a uniform magnetic field (B). One component of the motion of the cluster ions is at the cyclotron frequency (ω_c). The cyclotron resonance frequencies and hence the M/Q ratios (since $\omega_c = QB/M$) are determined by exciting the cluster ion cloud with an electric pulse and then performing a Fourier analysis of the resulting ion cloud resonance. This type of mass spectroscopy is extremely accurate and mass resolutions of 10^{-6} have been measured.

1.3.4.6 Neutral Clusters

Neutral clusters are usually generated with a fairly wide distribution of sizes. Size selection of neutral clusters can be achieved in one of two ways (alternative approaches involving the deflection of clusters by inhomogeneous magnetic or electric fields do not provide good size resolution as the magnetic moments and polarizabilities of clusters generally vary slowly with cluster size). At present, truly single-sized ('monochromatic') neutral cluster beams are only possible for relatively small clusters (several atoms).

(a) Cluster beam deflection. Neutral atomic or molecular clusters can be separated according to size by colliding a cluster beam with a beam of rare gas atoms or ions. Due to the conservation of momentum, larger clusters are scattered by smaller angles than smaller clusters. In this way, the mass distribution of large nitrogen clusters, $(N_2)_N$ (with $N \approx 10,000$), has been determined and small argon clusters ($N \leq 10$) have been separated by making use of the different angular and velocity distributions of the different sized clusters after collision.

The main problem with this method of size selection is that the intensity of the scattered neutral cluster beam is usually low. In order to minimize collisional dissociation, the collision energy is required to be small and the mass of the scattering atom should also be small. For size separation to be observed, an intense, high resolution cluster beam is required, with expansion conditions that minimize the angular ($\Delta\theta < 0.5°$) and velocity spread ($\Delta v/v < 0.05$) of the beam. The higher the mass and velocity of the clusters, the lower the resolution of the size selection. By contrast, the higher the mass and velocity of the scattering atoms, the higher the resolution—though, as discussed above, this increases the likelihood of cluster fragmentation.

(b) Re-neutralization of ions. A commonly used method for generating size-selected beams of neutral clusters is to generate positively or negatively charged clusters (as described earlier), size-select them by deflection in a magnetic or electric field and then re-neutralize the cluster ions. It is important, however, that the re-neutralization step should not cause excessive cluster fragmentation, or the size distribution of the neutral beam will not reflect that of the ions.

Cluster anions can be re-neutralized by photodetachment:

$$A_N^- + h\nu \rightarrow A_N + e^-$$

(though this process is known to cause significant fragmentation in the case of silver clusters) or by collisional electron detachment or charge exchange with a more electronegative element:

$$A_N^- + B \rightarrow A_N + B + e^-$$
$$A_N^- + X \rightarrow A_N + X^-$$

though collision-induced fragmentation may be a problem here.

Cluster cations are re-neutralized by charge exchange with a more electropositive element (e.g. an alkali metal atom):

$$A_N^+ + M \rightarrow A_N + M^+$$

Again collision-induced dissociation is a competing reaction. The higher the collision energy, the greater the amount of cluster fragmentation. It has also been found, however, that when the electron transfer process is strongly exothermic the neutral cluster is generated in an excited state and readily undergoes fragmentation. A detailed study has been made of the re-neutralization of sodium clusters by caesium atoms:

$$Na_N^+ + Cs \rightarrow Na_N + Cs^+$$

where it was found that the main products of the re-neutralization were Na_N, Na_{N-1} and Na_{N-2}. For Na_7^+, the electron transfer reaction with Cs is exothermic by a small amount (0.13 eV) so the cross-section for electron transfer is high, while that for fragmentation is small.

The advantages of the re-neutralization approach over cluster beam deflection are that it can be applied to all cluster sizes and it produces more intense neutral beams. Its main disadvantages, however, are that it generates neutral clusters with higher velocities and possibly with high internal excitation.

1.3.5 Investigation of Clusters

In this section, a brief overview is presented of a number of types of cluster experiment. More details of some of these techniques can be found in subsequent chapters, along with discussion and interpretation of the results of their application to specific clusters.

1.3.5.1 Media for Cluster Experiments

There are four main media for studying the nature (geometric and electronic structure and physical and chemical properties) of clusters: molecular beams; inert matrices; supported on surfaces; and in the solid state.

(a) Cluster molecular beams afford the opportunity to study isolated clusters, free from the influence of ligands or supports. While this is clearly desirable, and molecular beam technology has developed tremendously over the past decade, such studies of free clusters present difficulties associated with generating a sufficiently high flux of clusters with a narrow enough size distribution to give definitive information on specific clusters.

(b) Matrix isolated clusters. Clusters can be deposited in an inert matrix, which may be liquid, glassy or crystalline, and is generally made up of condensed rare gases or molecules (e.g. N_2 or CH_4). Matrix isolated clusters may be studied by direct UV-visible and IR spectroscopy, electron spin resonance and other spectroscopic techniques.

(c) Surface-supported clusters. When (as described in Section 8.4) clusters are supported on the surface of an inert substrate, such as an ionic oxide or a layered semiconductor (e.g. MoS_2 or graphite) it opens up the possibility of studying individual clusters using surface microscopy techniques. These include scanning tunnelling microscopy, scanning electron microscopy (SEM) and atomic force microscopy (AFM). For larger particles, X-ray and electron diffraction techniques can be used to study the degree of crystalline order in individual clusters.

(d) Cluster solids. In recent years, it has proved possible to crystallize solids composed of clusters. In order to prevent cluster coalescence the clusters are generally coated by surfactant molecules—usually organic thiols or thioethers, as described in Section 8.2. These cluster crystals may be studied using X-ray diffraction and electron microscopy.

1.3.5.2 Mass Spectroscopy: Factors Influencing Mass SpectralIntensities

The first experiments that were performed on clusters were mass spectroscopic in nature and these experiments remain important in cluster science, as they provide information (direct or indirect) on the sizes and (*vide infra*) stabilities of clusters. Most cluster mass determination methods involve the deflection of charged (generally positive) cluster cations by a magnetic or electric field. Methods for generating, mass selecting and detecting cluster ions have been discussed above. Here, I will concentrate on the factors which influence the measured intensities of peaks in the mass spectra of cluster beams.

Interpretation of mass spectral intensities and magic numbers, in terms of cluster stability, is not straightforward. For a given cluster size (N), the measured ion signal (MS intensity, I_N) can be written as a function of a number of factors which depend on the type of cluster and the experimental set up:

$$I_N = F\left(P_N, S_N, X_N, D_N\right) \tag{1.14}$$

(a) Cluster production efficiency (P_N). As mentioned above, the distribution of neutral cluster sizes depends on the type of cluster source being used and the operating conditions (p_0, T_0, A), which influence the relative rates of cluster nucleation, growth and evaporation. The proportion of cluster material to inert carrier gas is also important. For supersonic jet cluster sources, after leaving the nozzle clusters of a given size will persist if the adiabatically expanding carrier gas provides sufficient cooling to prevent further evaporation. The cluster production source may itself generate ions.

(b) Cluster stability (S_N). If there is an inherent stability associated with a given number of atoms in a neutral cluster then, all other factors being equal, this will give rise to a greater abundance of this clusters and a large peak in MS intensity (i.e. a 'magic number'), relative to similarly sized clusters. How the increased abundance of the more stable clusters comes about depends on the operating conditions (mentioned above) in the source and the nozzle conditions.

If the abundance is dominated by evaporative cooling, then, applying the evaporative ensemble model of Bjørnholm and Klots, the ratio of adjacent peak intensities is given by:

$$\ln(I_N / I_{N+1}) = \frac{D_2 A(N)}{kT} \qquad (1.15)$$

where $D_2A(N)$ is the second difference in Helmholtz free energy ($A = U - TS$):

$$D_2 A(N) = 2A(N) - A(N-1) - A(N+1) \qquad (1.16)$$

An identical distribution is predicted by the quasi-equilibrium model of de Heer and co-workers, in which the cluster abundance reflects the quasi-equilibrium distribution in the nozzle region, which is rapidly frozen by the adiabatic supersonic expansion. In this model, the relative peak intensities:

$$\ln(I_N / I_{N+1}) = \frac{D_2 E_b(N)}{kT} \qquad (1.17)$$

depend on the second difference in the binding (or cohesive) energies (E_b) of the clusters:

$$D_2 E_b(N) = 2E_b(N) - E_b(N-1) - E_b(N+1) \qquad (1.18)$$

Magic numbers are found to be most prominent when cluster generation leads to establishment of a quasi-equilibrium, due to high initial cluster temperatures (e.g. 300–600 K for alkali metal clusters) and slow cooling. On the other hand, Whetten and co-workers have shown that, for alkali metal clusters, kinetically limited sources, involving low-temperature (< 100 K) laser ablation in cold flowing helium and rapid cooling (by supersonic expansion) generate a cluster beam which is far from equilibrium and for which the MS intensities do not show any magic numbers. It was observed, however, that if the flux of the ionizing laser beam was increased, magic numbers were again observed. This is because at high laser flux multi-photon fragmentation occurs and the observed intensities either reflect clusters which are more stable to fragmentation or which are the more stable fragmentation products.

(c) Ionization efficiency (X_N). As mass spectroscopy depends on the detection of charged ions—normally cations in the case of metal clusters—care must be taken when inferring the relative abundance (and hence considering the relative stabilities) of neutral metal clusters from the measured abundance of cationic clusters. The ionization efficiency term is a function both of the ionization probability (the ionization cross section) and the likelihood of fragmentation accompanying the ionization process. Thus, for a measured MS intensity distribution of M_N^+ cations to reflect the relative abundance of neutral clusters, M_N prior to ionization requires:

- ionization cross sections which are approximately constant (or smoothly varying) as a function of N,

- low fragmentation cross sections of the initially generated ions.

Ion fragmentation is less likely when low energy ionization, in particular photoionization, is performed. However, if the photon energy is close to the ionization energy of the clusters (near-threshold ionization) then the ionization cross-section efficiency can vary dramatically as a function of size. This is because, for low photon energies, the cross section is inversely related to the cluster ionization energy. Since the most stable clusters generally have higher ionization energies than their neighbours, near-threshold ionization can lead to a trough (or dip), rather than a peak, in the MS intensity.

(d) Detection efficiency (D_N). This factor includes contributions from the focussing efficiency of the ion source and efficiency of mass selection by the mass filter. For most detectors and cluster types, the efficiency of detection of the ionized clusters varies smoothly and monotonically across the size range.

1.3.5.3 Other Types of Experiment

(a) Spectroscopy. In the case of small metal clusters, it is possible to generate high intensity molecular beams of size-selected clusters. In this case the measurement of direct UV-visible and IR absorption spectra are possible. However, experiments on free clusters in molecular beams generally make use of the high sensitivity of mass spectroscopy for the detection of charged particles. Thus, a UV-visible absorption spectrum of a beam of mass selected ionized clusters can be measured by scanning the frequency of the laser used for photoexcitation. At a frequency corresponding to an allowed optical transition in the cluster, the laser pumps sufficient energy into that absorption to dissociate the cluster. The absorption spectrum is traced out by monitoring the intensity of the mass of the undissociated cluster as a function of laser frequency. This process is known as *depletion spectroscopy* (or *photodepletion spectroscopy*). Depletion spectroscopy is more useful than direct absorption spectroscopy for low concentrations of clusters because of the high sensitivity of mass spectroscopy, which can detect individual charged clusters.

(b) Electron removal. As mentioned above, the high sensitivity of detectors for charged particles means that experiments involving charge separation are particularly useful for clusters. Such experiments include the measurement of *photoelectron spectra* of neutral or anionic clusters, as well as the measurement of traditional mass spectral abundances of cationic or anionic clusters. Magnetic or electric fields can be used to deflect (or collect) cluster cations or anions (depending on their charge/mass ratio) or the ejected electrons (according to their kinetic energy).

(c) Cluster fragmentation. Neutral and ionized clusters can be excited either by absorption of light (as in spectroscopic studies) or by electron or ion impact. The excited clusters may then relax by emitting radiation, losing an electron and/or evaporation/fragmentation. The radiation or particles emitted can then be detected. *Photofragmentation* (photodissociation) has been discussed above in the context of depletion spectroscopy, but it can also give information as to the cluster binding energy, as can the related *collision-induced fragmentation/dissociation* process.

(d) Polarizability and magnetism measurements. Electronic polarizabilities are measured by passing a collimated beam of neutral clusters through an inhomogeneous electric field, with the deflection of the clusters being proportional to the cluster polarizability. The magnetism of metal clusters can be measured in an analogous way, by deflecting the cluster beam in an inhomogeneous magnetic field (i.e. a Stern–Gerlach type of experiment). Electron spin (paramagnetic) resonance experiments have also been performed on small metal clusters supported in inert gas matrices.

(e) Cluster ion mobility studies. Bowers and Jarrold have pioneered the development of cluster ion chromatography, for metal and semi-conductor clusters, based on cluster ion mobility. In these experiments, cluster ions (generally produced in a laser vaporization source) are mass selected and injected into a long drift tube (generally with 50 μm inlet and outlet diameters), which is filled with an inert buffer gas (usually helium) at a controlled temperature and pressure. Interestingly, ion mobility spectrometers (where the ion mobilities are measured in air) are currently being used by the US army and NATO as portable detectors for chemical warfare agents.

The cluster mobilities (which are inversely related to the time taken to pass through the drift tube) depend on the number of collisions with the buffer gas and these in turn depend on the collisional cross sectional area, and hence the shape, of the cluster. Assuming that the collisions with the buffer gas do not lead to preferential orientation of the clusters, for a given number of atoms (and density), spherical clusters have the smallest collision cross sections and therefore travel fastest through the drift tube. As they rotate in the drift tube, prolate spheroidal clusters (where one dimension or axis of the cluster is appreciably longer than the other two) carve out a large sphere, and thus have high collision cross sectional areas and slower drift times. Oblate, clusters (where one dimension or axis of the cluster is appreciably shorter than the other two) have collisional cross-sectional areas, and hence drift times, intermediate between those of spherical and prolate geometries. In this way cluster isomers, which have the same number of atoms (and hence the same mass) but different shapes, are temporally separated in the drift tube and appear at different times at the detector.

(f) Flow reactor studies. In a flow reactor, such as that developed by Riley and co-workers, clusters are pre-formed using one of the methods described above and then passed through a second 'pick-up' chamber at a pressure of around 10^{-6} Pa where molecules (such as H_2, N_2 and CO) adsorb onto the cluster surface. As will be shown later (Section 5.2) flow reactor experiments have been used to study the relative uptake of small molecules and relative reactivities of metal clusters as a function of cluster size and elemental composition.

(g) Diffraction experiments. *Electron diffraction* studies have been performed on rare gas and metal clusters in cluster molecular beams. In such experiments, a well collimated electron beam (with electron energies in the range 30–50 keV) is crossed with a cluster beam. The electrons are scattered by atoms in the cluster and the diffraction pattern (arising from interference of the scattered electrons) is collected on a photographic film or by an electron detector. Because the

information from diffraction experiments is averaged over all clusters that the electrons interact with, electron diffraction measurements require a narrow cluster size distribution and a reasonably high cluster intensity. A number of *X-ray diffraction* studies have also been carried out on metallic clusters deposited on inert substrates.

In both electron and X-ray diffraction studies of clusters, the diffraction pattern is interpreted by generating a model of the cluster and adjusting the model so as to maximize the agreement with the experimentally measured pattern. Diffraction experiments can be used to determine: the structures, sizes (from a Scherrer analysis of the peak widths—with smaller clusters giving rise to broader peaks) and mean temperatures of the clusters.

(h) Microscopy. The most direct way of determining the structure of a cluster (or at least of its outer layers) is by using microscopy. This necessitates the immobilization of the cluster on a substrate (see Chapter 8).

Because of the small length scales (of the order of 10^{-10} m) of atomic diameters and interatomic distances, traditional optical microscopy (i.e. using visible light) cannot be used. Instead, the atomic structure of clusters is imaged by electron microscopy, using electron beams, that can be accelerated to an appropriate energy (and hence wavelength) and can be focussed by electrostatic lenses. Electron microscopy techniques, such as *transmission electron microscopy* (TEM), *scanning electron microscopy* (SEM) and *high resolution electron microscopy* (HREM) can now achieve atomic resolution.

In *scanning probe microscopy* (SPM) techniques, the shape (topology) of surfaces and clusters are mapped out using a needle tip that is positioned by piezo-electric drives, with an accuracy of 10^{-11} m (i.e. smaller than the width of a single atom). This enables such techniques to display atomic resolution. The needle tip is tracked backwards and forwards (scanned) across the substrate and detects atoms in one of two ways. In *scanning tunnelling microscopy* (STM), a potential bias is applied between the needle tip and the substrate, causing electrons to tunnel from the surface to the needle (or the reverse, depending on the sign of the bias potential). The STM can operate in constant current mode (where the height of the tip above the substrate is varied so as to keep the tunnelling current constant) or in constant height mode (where the varying tunnelling current is measured). Another variant of scanning probe microscopy is *atomic force microscopy* (AFM), where the repulsive force between the tip and the substrate is measured.

1.4 EXERCISES

1.1 Calculate how large a spherical cluster must be before the number of surface atoms drops below 10% of the total number of atoms in the cluster.

1.2 As mentioned in Section 1.2.1, the Spherical Cluster Approximation (SCA) fails to take into account the packing fraction (f—i.e. the extent to which hard spheres can pack together to fill 3-D space). For a given total number of atoms (N), what is the qualitative effect on the number of surface atoms

(N_s) of having a packing fraction $f < 1$? Derive a new expression for the fraction of surface atoms in a pseudo-spherical cluster with fcc packing, by including the packing fraction ($f_{fcc} = 0.74$), while keeping all the other approximations of the SCA. (See Equation (4.31), if help is needed.)

1.3 List the experimental factors that influence the measured peak intensities in the mass spectrum of a beam of clusters.

1.4 Describe methods for preparing gas phase clusters of a rare gas (e.g. argon) and a refractory metal (e.g. iron).

1.5 FURTHER READING

Attfield, J.P., Johnston, R.L., Kroto, H.W. and Prassides, K., 1999, New science from new materials. In *The Age of the Molecule*, edited by Hall, N., (London, Royal Society of Chemistry), pp. 181–208.

Berry, R.S., 1994, Melting and freezing of clusters: how they happen and what they mean. In *Clusters of Atoms and Molecules*, Vol.I, edited by Haberland, H., (Berlin, Springer), pp. 187–205.

Drexler, K.E., 1990, *The Engines of Creation*, (London, Fourth Estate).

Duncan, M.A. and Rouvray, D.H., 1989, Microclusters. *Scientific American* (December 1989), pp. 60–65.

Echt, O., 1996, Convergence of cluster properties towards bulk behavior: how large is large? In *Large Clusters of Atoms and Molecules*, edited by Martin, T.P., (Dordrecht, Kluwer), pp. 221–239.

Haberland, H., 1994, Experimental methods. In *Clusters of Atoms and Molecules*, Vol.I, edited by Haberland, H., (Berlin, Springer), pp. 207–252.

de Heer, W. A., 1993, The physics of simple metal clusters: experimental aspects and simple models. *Reviews of Modern Physics*, **65**, pp. 611–676.

Johnston, R.L., 1998, The development of metallic bonding in clusters. *Philosophical Transactions of the Royal Society of London*, **A356**, pp. 211–230.

Jortner, J., 1992, Cluster size effects. *Zeitschrift für Physik*, **D24**, pp. 247–275.

Kappes, M.M. and Leutwyler, S., 1988, Molecular Beams of Clusters. In *Atomic and Molecular Beam Methods*, Vol. 1, edited by Scoles, G., (Oxford: Oxford University Press), pp. 380–415.

Müller, H., Fritsche, H.-G. and Skala, L., 1994, Analytical cluster models and interpolation formulae for cluster properties. In *Clusters of Atoms and Molecules*, Vol.I, edited by Haberland, H., (Berlin: Springer), pp. 114–140.

Shvartsburg, A.A., Hudgins, R.R., Dugourd, P. and Jarrold, M.F., 2001, Structural information from ion mobility measurements: applications to semiconductor clusters. *Chemical Society Reviews*, **30**, pp. 26–35.

Whetten, R.L. and Schriver, K.E., 1989, Atomic clusters in the gas phase. In *Gas Phase Inorganic Chemistry*, edited by Russell, D.H., (New York: Plenum), pp. 193–226.

CHAPTER TWO

Rare Gas Clusters

2.1 INTRODUCTION

Rare gas clusters were the earliest clusters to be investigated in cluster molecular beam experiments and they remain the most widely studied of clusters—because of the ease with which they can be generated, the low melting and boiling points of the rare gas elements and their low reactivity. The bonding in rare gas clusters is simple—to a first approximation cluster binding energies are pair-wise additive, but clusters of the lightest rare gas element, helium, exhibit quantum effects, such as superfluidity and will be discussed separately in Section 2.5.

2.1.1 Nomenclature

The rare gases are the elements of group 18 of the periodic table: helium (He), neon (Ne), argon (Ar), krypton (Kr), xenon (Xe) and radon (Rn). The name 'rare gas' is an historical term, because at the time of their discovery, at the end of the nineteenth century, they were believed to be very rare. Although relatively rare on earth (arising from α-particle emission) helium is actually the second must abundant element in the Universe, so its definition as a rare gas is somewhat misleading. Argon is more abundant terrestrially, constituting 1% of the atmosphere. The radioactive gas radon is found in relatively high abundance (and can lead to significant health risks) in buildings constructed from granite, as radon is a decay product of the uranium contained in the granite.

Another old name for this group of elements is the 'inert gases' since it was once believed that they were chemically inert, never forming compounds with other elements. Although helium, neon and argon do not undergo chemical reactions, consistent with their closed electronic shell nature (see below), the heavier elements krypton and xenon (and presumably radon), which have lower ionization energies than their lighter congeners, do form compounds with electronegative elements, such as oxygen and fluorine (e.g. XeF_6).

A third collective name for these elements, which has been introduced more recently is 'noble gases' to indicate the fact that they rarely react. However, as in most physics texts the old name of 'rare gases' still prevails, I will use this name to describe this group of elements. I will also use the symbol Rg to refer to a generic rare gas element.

2.1.2 Bonding in Rare Gas Clusters

The rare gas elements, which lie on the extreme right-hand side of the periodic table of the elements, have closed shell electronic configurations. The electronic configuration of helium is $(1s)^2$, while the heavier rare gas elements have outer (valence) shell configurations $(ns)^2(ns)^6$, where $n = 2$ for Ne, 3 for Ar, 4 for Kr, 5 for Xe and 6 for Rn. It is this closed shell nature, together with the high atomic ionization energies (arising from high effective nuclear charges), which render the rare gases so chemically inert (or noble).

The closed shell electronic configuration of the rare gas atoms also results in weak interatomic interactions between rare gas atoms. As shown schematically in Figure 2.1, the helium dimer has no net-covalent bonding, because both the bonding (σ_g) and the antibonding (σ_u^*) molecular orbitals, arising from the overlap of the atomic $1s$ orbitals, are doubly occupied, thereby cancelling each other out. For the heavier rare gas dimers, all the bonding and antibonding molecular orbitals, arising from overlap of the valence ns and np atomic orbitals, are occupied, again resulting in no net covalent bonding.

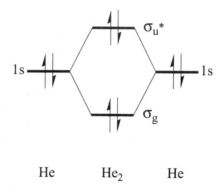

Figure 2.1 Molecular orbital diagram for the He₂ dimer.

2.1.2.1 Dispersion Forces

In the absence of covalent bonding, the bonding interaction between rare gas atoms is weak and is dominated by (London) dispersion forces. This attractive interaction is caused by fluctuations in electron density, which give rise to instantaneous electronic dipoles (and higher multipoles), which in turn induce dipoles in neighbouring atoms. Long range attractive dispersion forces, which are the weakest of the non-covalent or van der Waals (vdW) forces, therefore arise from dynamic electron correlation. A general expansion for the attractive energy (V_{disp}) due to these dispersion forces can be written:

$$V_{\text{disp}}(r) = -\left\{ \frac{C_6}{r^6} + \frac{C_8}{r^8} + \frac{C_{10}}{r^{10}} + \ldots\ldots \right\}$$

(2.1)

where the terms C_n are constants and r is the interatomic distance. The first term represents the instantaneous dipole–dipole interaction and is dominant, so the higher terms are often omitted when calculating the vdW forces between rare gas atoms. A reasonable approximation to the dispersion energy is given by the London formula (derived by London, based on a model proposed by Drude):

$$V_{disp} = -\frac{C_6}{r^6}, \quad C_6 = \frac{3\alpha^2 I}{4(4\pi\varepsilon_0)^2} \tag{2.2}$$

where α and I are the atomic polarizability and ionization energy, respectively.

There is a limit to the compressibility of matter, due to repulsive interactions which dominate at short range. At short internuclear separations there are electrostatic repulsions between the atomic nuclei, as well as between the electrons (both valence and core) on neighbouring atoms. In addition to these electrostatic repulsions there is also a short range Pauli repulsion between the electrons on neighbouring atoms, which is quantum mechanical in origin and derives from the Pauli exclusion principle. The total short range repulsive interaction is generally modelled by a steep $1/r^n$ (where n is usually 12) dependence on interatomic distance, though an exponential ($e^{-\alpha r}$) dependence is probably more accurate.

Combining the leading term in the attractive dispersion series with a $1/r^{12}$ repulsive term leads to the famous Lennard–Jones (LJ) model pair potential energy function, which has been used successfully to model rare gas dimers, clusters and condensed phases. For a pair of rare gas atoms (i and j) separated by a distance r_{ij}, the LJ potential may be written as:

$$V_{ij}^{LJ}(r_{ij}) = \varepsilon \left\{ \left(\frac{r_0}{r_{ij}}\right)^{12} - 2\left(\frac{r_0}{r_{ij}}\right)^6 \right\} \tag{2.3}$$

where ε is the well depth (the binding energy of the dimer) and r_0 is the equilibrium internuclear separation.

2.1.2.2 Binding Energies of Rare Gas Dimers

Values of the LJ well depth, ε, are listed in Table 2.1 for a number of rare gas dimers. Also listed are the corresponding temperatures (T_d, from $\varepsilon = kT_d$)—i.e. the temperature below which dimers can form. The increase in ε, and hence in T_d, on going to heavier rare gases reflects the increase in the attractive dispersion forces because of the higher polarizability and lower ionization potential, due to the lower effective nuclear charge experienced by the outer electrons in the heavier rare gas atoms. Because of the weak nature of vdW forces, the binding energies of all the rare gas dimers are small, the maximum being only 24 meV for Xe_2. This should be contrasted with the corresponding binding energy of the hydrogen molecule (which has a single covalent bond) of 4.8 eV—i.e. over two orders of magnitude greater. The special case of helium clusters (where vibrational energies are critically important) will be discussed in Section 2.5.

As shown in Table 2.1, the boiling temperatures (T_b) of the rare gas elements follows the same trend as that observed for the ε and T_d values of the dimers (i.e. increasing with increasing atomic number) because of the increasing dispersion forces. Finally, Table 2.1 shows that the same trend is observed for the melting temperatures (T_m) of the rare gas elements. In fact, helium cannot be solidified at pressures lower than 25 atmospheres.

2.1.2.3 Many-Body Forces

On moving from dimers to larger clusters, or to the bulk solid, accurate modelling of the interatomic forces between the rare gases requires the addition of non-pairwise additive many-body forces, since it is known that even for a cluster with three atoms, the total potential energy is not just the sum of the three pair interactions $(V_{abc} \neq V_{ab} + V_{bc} + V_{ac})$. This non-additivity reflects the fact that the dispersion attraction between two atoms is modified by the presence of other neighbouring atoms. Many-body effects, however, are generally small, accounting for only 10% of the lattice energy of solid argon, for example. For this reason, most simulation studies of rare gas clusters, liquids and solids have been performed using the pairwise additive Lennard–Jones potential.

Table 2.1 Dimer well depth (ε), dimerization temperature (T_d), boiling point (T_b) and melting point (T_m) for rare gas elements.

Element	ε / meV	T_d / K	T_b / K	T_m / K
He	0.9	11	4.2	0.95[*]
Ne	4	42	27.1	24.6
Ar	12	142	87.3	83.8
Kr	17	200	120	116
Xe	24	281	165	161
Rn	—	—	211	202

[*] $P = 26$ atmospheres

2.2 MASS SPECTRAL STUDIES OF RARE GAS CLUSTERS

2.2.1 Generation and Growth of Rare Gas Clusters

Rare gas clusters are generated by supersonic expansion from a high-pressure region into a vacuum, through a narrow nozzle, as described in Chapter 1. A large number of collisions occur during the initial expansion phase, giving rise to clusters. The supersonic expansion leads to low relative kinetic energies and low cluster temperatures. If the temperature is less than T_d (i.e. $T < \varepsilon/k$) then dimers can form, though three-body collisions are still required in order to remove excess kinetic energy. The Rg_2 dimers act as nucleation sites for further cluster growth, via condensation and cluster collision. Cluster condensation leads to cluster heating, which can eventually lead to evaporation of atoms from the cluster, thereby removing some of the internal energy as kinetic energy of the evaporated atoms, while leaving the clusters vibrationally and rotationally hot. Thus, there is a competition between growth and evaporation, in the so-called evaporative ensemble discussed in Chapter 1.

2.2.2 Cluster Ionization

In mass spectrometric studies of rare gas clusters, the clusters are detected as cations, which are generated either by electron bombardment or photoionization. The ionization process generally imparts 1–2 eV of energy to the cluster, which leads to cluster heating and partial evaporation. Thus, the abundances observed in mass spectroscopic studies of rare gas clusters reflect the stabilities of the ionized clusters towards evaporation.

2.2.3 Mass Spectra of Rare Gas Clusters

The mass spectrum of xenon clusters, Xe_N ($N \leq 150$), was measured by Echt and co-workers in 1981. In the early 1980s, spectra were also measured for He_N ($N \leq 32$) by Stephens and King, and for Ar_N and Kr_N ($N \leq 60$) by Ding and Hesslich. Among other, more recent work, it is worth noting the studies by Märk and co-workers on Ne_N ($N \leq 90$) and Rg_N (Rg = Ar, Kr and Xe; $N \leq 1,000$). Friedman and Buehler also found that the growth of large argon clusters (with an average N of approximately 560) was promoted by using a small concentration (2%) of Ar^+ ions as seeds for cluster nucleation, with a large excess of He gas to cool the growing clusters.

Figure 2.2 shows the mass spectra of Ar, Kr and Xe clusters with up to 150 atoms. The figure shows that the intensities of the cluster peaks are not smoothly varying with cluster size. There are a number of relatively intense peaks and other cases where an abrupt drop in intensity is observed. This pattern is particularly evident in the mass spectrum of xenon clusters (Figure 2.2c). The nuclearities corresponding to intense MS peaks are termed 'magic numbers', by analogy with the magic numbers of nucleons observed in nuclear physics.

2.2.4 Magic Numbers and Structures of Rare Gas Clusters

Depending on the experimental conditions (as discussed in Chapter 1), the magic numbers observed in the mass spectra of rare gas clusters may arise due to the size-dependence of the binding energy of the Rg_N^+ cations and the fragmentation (i.e. evaporation) process that occurs after ionization. It is therefore difficult to infer directly the abundance of the initially generated neutral rare gas clusters from the measured abundance of cationic clusters. In particular, as will be discussed in Section 2.3, the electronic structure of Rg_N^+ is quite different from that of the neutral cluster and there is still some debate as to whether neutral and cationic clusters have the same geometries.

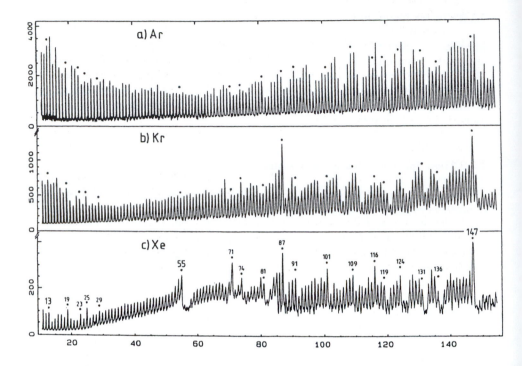

Figure 2.2 Mass spectra of positively charged Ar, Kr and Xe clusters. The distinct intensity anomalies for Xe clusters are marked by dots and the magic numbers of atoms are given.

Inspection of the mass spectra shown in Figure 2.2 (especially for Xe clusters) reveals local intensity maxima at the following magic number nuclearities: $N^* = 13, 19, 25, 55, 71, 87$ and 147. Smaller features (secondary magic numbers) are observed at $N^* = 23, 81, 101$ and 135. Several of these magic numbers may be rationalized in terms of cluster structures consisting of concentric geometric (polyhedral) shells of atoms around a central atom. The best agreement with experiment is for structures composed of icosahedral shells, with multiple five-fold symmetry axes. It is easily shown that, for a geometric shell cluster composed of K complete icosahedral shells, magic numbers are given by:

$$N^*(K) = 1 + \sum_{k=1}^{K}\left(10k^2 + 2\right) \tag{2.4}$$

which can be expanded to give:

$$N^*(K) = \frac{1}{3}\left(10K^3 + 15K^2 + 11K + 3\right) \tag{2.5}$$

This explains the peaks at $N^*=13$ ($K=1$), $N^*=55$ ($K=2$) and $N^*=147$ ($K=3$). The next magic numbers in this sequence are $N^*=309$ ($K=4$) and $N^*=561$ ($K=5$). The outer shells of these icosahedral clusters (known as Mackay icosahedra) are shown in Figure 2.3a. Geometric shells will be discussed at greater length in Chapter 4, in the context of the structures of metal clusters.

These icosahedral shell magic numbers are consistent with calculations on neutral rare gas clusters, using model interatomic potentials (such as the Lennard–Jones potential), which predict a growth sequence based on maximizing the number of nearest-neighbour contacts, so as to maximize the total cluster binding energy. As shown in Figure 2.3b, this leads to an equilateral triangular geometry for $N = 3$ and a tetrahedron for $N = 4$. Compact larger clusters can be constructed by fusing tetrahedral units—so-called polytetrahedral growth—yielding, for example, the five-fold symmetric pentagonal bipyramid for $N = 7$ (constructed from five tetrahedra) and the centred (one-shell) icosahedron for $N = 13$ (twenty tetrahedra).

Some of the other magic numbers in the rare gas cluster mass spectra have been attributed to sub-shell closure, where an incomplete shell has been added to the previous icosahedral shell. The magic numbers $N^* = 19$ and 25 can be explained in terms of the fusing of two and three icosahedra respectively, along one of the five-fold axes of the icosahedron. These structures can alternatively be constructed by adding one or two six-atom caps to the Rg_{13} cluster. The double icosahedron structure proposed for Rg_{19} is shown in Figure 2.3c.

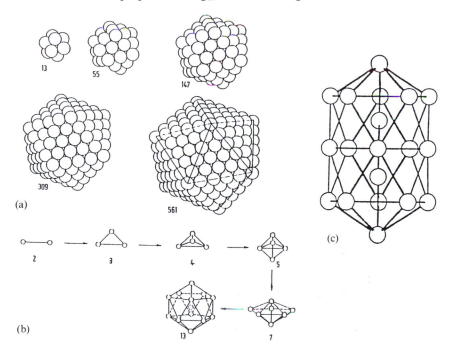

Figure 2.3 Structures of rare gas clusters. (a) Icosahedral geometric shell clusters with 13–561 atoms. (b) Polytetrahedral growth sequence of small clusters. (c) The double icosahedral structure of Rg_{19}.

Theoretical calculations have shown that, starting from a complete Mackay icosahedron, additional atoms are initially added to the highest coordination sites—on the faces of the icoshedron, such that each added atom sits on a three-fold hollow on the surface of the parent icosahedron (thereby maximizing it's coordination). Continuation of this packing does not, however, lead to the next Mackay icosahedron (in fact it is known as anti-Mackay packing). The next shell of the Mackay icosahedron has atoms sitting above pairs of atoms on the surface of the parent icosahedron. As more atoms are added to the incomplete shell, eventually the Mackay packing becomes favoured over anti-Mackay, as more atoms can be accommodated in the Mackay icosahedron and this also maximizes the number of interatomic interactions within the growing shell.

The calculated sublimation energies (the energy required to remove an atom from an N-atom cluster) for argon clusters, with up to 45 atoms, are shown in Figure 2.4. The sublimation energy of an N-atom cluster, ΔE_{sub}, is calculated as the difference in binding energy (E_b) of the Rg_N and Rg_{N-1} clusters:

$$\Delta E_{sub}(N) = E_b(N) - E_b(N-1) \tag{2.6}$$

For a pairwise additive potential energy function, such as the Lennard-Jones potential, the binding energy (a positive quantity) is merely the negative of the sum of all interactions, V_{ij}, between pairs of atoms (i and j):

$$E_b(N) = -\sum_{i=1}^{N-1} \sum_{j=i+1}^{N} V_{ij} \tag{2.7}$$

The high sublimation energies shown in Figure 2.4, for nuclearities corresponding to closed shell and closed sub-shell structures, confirms the high relative stabilities of these structures. Also noticeable is the gradual increase in sublimation with increasing cluster size, as for larger clusters a greater number of interactions must be broken in order to remove an atom. Thus, while the Ar_2 dimer has a binding energy $E_b(2) = \varepsilon = 12$ meV (see Table 2.1), bulk argon has a sublimation energy of 80 meV and magic number clusters Ar_{13} and Ar_{19} have intermediate sublimation energies of approximately 60 meV.

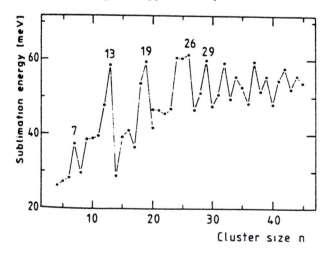

Figure 2.4 Sublimation energies calculated for Ar clusters.

2.2.5 Icosahedral vs. fcc Cluster Growth

2.2.5.1 Theoretical Considerations

As discussed in the previous section, polytetrahedral growth (see Figure 2.3a), leading eventually to icosahedral geometric shell clusters, lowers the cluster energy by maximizing the number of near-neighbour attractive interactions. As can be seen in Figure 2.3b, the icosahedral shell clusters have pseudo-close packed surfaces which resemble the close-packed (111) surfaces of a face-centred cubic (fcc) close packed solid. The surface coordination (number of nearest neighbours of surface atoms) and, therefore the overall number of nearest-neighbour interactions is greater for icosahedral than for other geometric shells, based on crystalline solid structures (e.g. fcc, hexagonal close-packed (hcp), or body-centred cubic (bcc) solids). As increased coordination generally equates to lower potential energy, the increased stability of icosahedral shell clusters over alternative geometric shell structures can be said to be driven by the lower surface energy of the icosahedral structures.

Although the minimization of the cluster's surface energy favours icosahedral structures, as the cluster gets larger there must, at some critical nuclearity, N_c, be a transition to the crystalline fcc structure, as the rare gas elements are known to adopt the fcc structure in the solid state. The five-fold symmetry of the icosahedron is incompatible with the three-dimensional periodicity required to form a commensurate crystal, but, in principle this could be accommodated by a small distortion of the cluster. More importantly however, the icosahedral cluster structures possess a bulk elastic strain which eventually destabilizes them relative to fcc-like structures.

As mentioned above, the thirteen-atom one-shell icosahedral cluster can be constructed from twenty tetrahedra. Figure 2.5a shows, however, that starting from twenty regular tetrahedra results in a structure with gaps between some of the tetrahedra. To eliminate the gaps and produce a regular icosahedron (Figure 2.5b), the tetrahedra have to be slightly distorted. Alternatively, if one constructs an icosahedron by sticking together thirteen hard spheres, as shown in Figure 2.5c, then although the outer twelve spheres are all in contact with the inner sphere, they are not in contact with each other. These surface gaps arise due to a frustration in the packing for icosahedral structures, since a regular icosahedron has tangential (surface) contacts which are approximately 6% longer than the radial contacts.

Icosahedral clusters generally lower their energy by contracting so that the surface distances are decreased, thereby getting closer to the minimum in the interatomic potential energy curve. This is achieved at the expense of making the radial distance a little too short, so that the attractive radial interactions are reduced. For small clusters, the lowering of the surface energy outweighs the effect of the shortened radial distances, but for larger clusters this core compression becomes larger and larger, building up a mechanical strain, which eventually destabilizes the icosahedral structure relative to fcc-like geometries. (See also the discussion in Section 4.6.1 for metal clusters.)

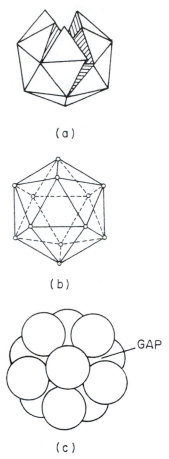

Figure 2.5 Representation of packing frustration in icosahedral Rg_{13} clusters. (a) Gaps left by twenty regular tetrahedra sharing a common vertex. (b) Regular icosahedron. (c) Icosahedral cluster of thirteen hard spheres, showing the gaps left by packing frustration.

2.2.5.2 Electron Diffraction Studies

Electron diffraction experiments by Farges and Lee (in the 1980s) on beams of argon clusters, confirmed that argon clusters of up to several hundreds of atoms are icosahedral, or are based on icosahedra. In an electron diffraction experiment, the coherent interference between elastically scattered electrons generates a snapshot of the cluster on a timescale of approximately 10^{-16} s.

In the experiments of Farges and Lee, size-dependent changes in the electron diffraction pattern were interpreted as being due to a change in structure from icosahedral to fcc-like (microcrystalline) at a critical nuclearity in the region $800 \leq N_c \leq 1,000$. Theoretical calculations (using for example the Lennard–Jones potential) lead to the prediction of much higher critical nuclearities ($N_c \approx 10,000$). One reason for this discrepancy may be that theoretical calculations are generally carried out at a cluster temperature of 0 K, while the cluster temperatures in the electron diffraction experiments was 38±4 K. Another possibility is that the high-energy (40–50 keV) electrons used in these diffraction experiments caused fragmentation of larger clusters, which indeed may have fcc structures, and which are responsible for the observed diffraction patterns.

For clusters with $N \lesssim 800$, the electron diffraction patterns have been interpreted in terms of multilayer (geometric shell) icosahedral structures, while smaller clusters (with up to 50–60 atoms) have the polytetrahedral structures, predicted by calculations using the LJ potential.

2.3 CHARGED RARE GAS CLUSTERS

In this section, the term 'charged cluster' will generally refer to positively charged clusters, since these are the most commonly studied. Negatively charged rare gas clusters will, however, be discussed in Section 2.3.6.

2.3.1 Bonding in Charged Rare Gas Dimers

There is a significant change in bonding upon removing an electron from a rare gas cluster. This can be best understood by considering the differences between the electronic structures of the neutral (Rg_2) and singly charged (Rg_2^+) dimers. As mentioned in Section 2.1.2, the He_2 dimer has the electronic configuration $(\sigma_g)^2(\sigma_u^*)^2$ and a covalent bond order of 0. Ionization of the cluster generates He_2^+, with electronic configuration $(\sigma_g)^2(\sigma_u^*)^1$. Since an electron has been removed from an antibonding orbital (see Figure 2.1), He_2^+ has a certain degree of covalent bonding (and a bond order of ½). This is reflected in a binding energy of 2.5 eV for He_2^+, compared with 1 meV for the neutral dimer.

For the heavier rare gas elements, the lowest energy electronic state ($^2\Sigma_u^+$) of Rg_2^+ is generated by removing an electron from the high-lying σ_u^* (np) antibonding orbital formed by out-of-phase combination of atomic np orbitals, oriented along the interatomic (bond) vector. Removal of this electron again leads to a formal covalent bond order of ½. In the case of argon, as shown in Figure 2.6, the binding energy of Ar_2^+ is 1.5 eV, compared with a value of 12 meV for the neutral dimer. The smaller increase in binding energy on ionization observed for argon, compared with helium, is because the argon $3p$ orbitals overlap less efficiently than the helium $1s$ orbitals.

As shown in Figure 2.6, the large increase in binding energy upon ionization of the rare gas dimer is accompanied by a decrease in the equilibrium Rg–Rg distance, reflecting the presence of partial covalent bonding. In the case of the argon dimer, there is a 30% decrease in the Ar–Ar bond length on going from Ar_2 to Ar_2^+. The stability of the Ar_2^+ ion is confirmed by mass spectral measurements (using a 70 eV electron beam for ionization) of a beam of neutral Ar_5 clusters (mass selected by momentum transfer from a beam of helium atoms). In this experiment, the Ar_2^+ peak is much larger than those due to Ar_3^+–Ar_5^+, indicating that Ar_2^+ is the dominant fragment channel when small argon clusters are ionized.

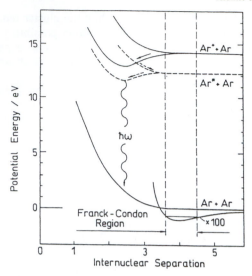

Figure 2.6 Potential energy curves for neutral, charged and excited Ar$_2$.

2.3.2 Charge Localization in Charged Rare Gas Clusters

Considering charged rare gas clusters of more than two atoms, one of the interesting questions concerns whether the positive charge is localized on a small subset of the atoms or delocalized over the entire cluster (as is the case for charged metal clusters). As indicated above, there is a large geometry change (i.e. a significant bond length reduction) in the rare gas dimer when it is ionized. If we consider the positive charge to initially be localized on a small number of atoms, it is reasonable to suppose that the movement of this positive charge over the entire cluster (i.e. electron-hole delocalization) should also be accompanied by cluster rearrangement. This rearrangement has an energy cost associated with it and will, thus, tend to oppose charge delocalization. This phenomenon is known as self-localization (or self-trapping) of charge and is the cluster counterpart of hole localization in rare gas solids.

Electronic structure calculations on positively charged neon clusters indicate that more than 97% of the charge is localized on a single pair of atoms. The charged neon cluster, Ne$_N^+$, can therefore be written as (Ne$_2^+$)Ne$_{N-2}$, consisting of a charged dimer core, surrounded by rings of neutral neon atoms. In the case of charged helium clusters, structures with charged dimer and trimer cores have nearly the same energy, while heavier charged rare gas clusters have ionic cores which may be dimeric, trimeric or even tetrameric, depending on cluster size and symmetry.

In general, charged rare gas clusters, Rg$_N^+$, consist of ionic cores (Rg$_C^+$) surrounded by ($N-C$) neutral Rg^0 atoms, which are polarized by the charge on the core. The outer atoms effectively solvate the charged core, as shown schematically in Figure 2.7.

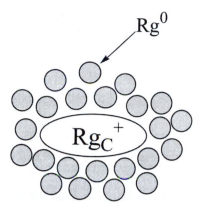

Figure 2.7 Schematic representation of a charged rare gas cluster.

The attractive induction energy of interaction between a polarizable Rg atom and an Rg^+ ion, separated by a distance r, is of the form:

$$V_{\text{ind}} = -\frac{\alpha e^2}{8\pi\varepsilon_0 r^4} \tag{2.8}$$

where α is the polarizability of the neutral Rg atom and e is the charge on the electron. As V_{ind} is larger than the attractive dispersion (van der Waals) forces found in neutral rare gas clusters, the energy required to eject a neutral atom from a charged cluster is greater than for a neutral cluster.

Photofragmentation experiments on charged argon clusters by Stace have shown that the Ar_3^+ and Ar_4^+ cores are linear, in agreement with theoretical calculations on argon and other rare gas clusters. Icosahedral Ar_{13}^+ clusters have linear Ar_3^+ cores, with charges of +0.5 on the central atom and +0.25 on the two outer atoms, while double icosahedral Ar_{19}^+ clusters have linear Ar_4^+ cores, with charges of +0.4 on the two inner atoms and +0.10 on the two outer atoms. Similar results apply for the other rare gas elements.

2.3.3 Generation and Stability of Charged Rare Gas Clusters

Threshold ionization energies can be measured by performing a photoelectron experiment, wherein synchrotron radiation is used to ionize a beam of clusters and only those photoelectrons with kinetic energies close to zero are collected. By collecting the corresponding cluster (photo-)cations at the same time, a coincidence experiment is carried out, which enables the determination of the energy required for each cluster to be ionized.

There is a significant decrease in ionization energy (typically of 1–2 eV) on going from a single rare gas atom to the dimer because of the stabilization of the Rg_2^+ cation by partial covalent bond formation. For larger clusters, the ionization energies decrease more gradually with increasing cluster size, towards the bulk limit. For argon clusters, the ionization energy of Ar_{20} is close to the bulk limit, while for krypton, Kr_{20} has an ionization energy which is still over 0.5 eV higher than the bulk ionization energy.

It is interesting to note that adiabatic ionization energies can be measured for rare gas clusters, since it is known that electronic excitation (at photon energies below the ionization limit) leads to cluster fragmentation (see Section 2.4). Because of the large difference in equilibrium bond length between neutral and charged rare gas dimers (Figure 2.6), the Franck–Condon factor for the direct photoionization process, $Rg_2 + h\nu \rightarrow Rg_2^+ + e^-$, is negligibly small. Ionization of rare gas dimers, and larger clusters, is believed to occur via the formation of an intermediate highly excited Rydberg state of the cluster (see Section 2.4), which subsequently undergoes structural relaxation and autoionization. After ionization, there is usually a large amount of excess energy (1–2 eV) in the cluster ion as the charge becomes localized (trapped) on a dimer, trimer or tetramer core unit (Rg_C^+) which is generated in a highly excited vibrational state (due to the difference between the equilibrium internuclear separations between the neutral and charged cores). Vibrational relaxation of the charged core is accomplished by evaporation of neutral rare gas atoms. The overall ionization process can therefore be summarized as:

$$Rg_N + h\nu \rightarrow (Rg_N^+)^* + e^- \rightarrow (Rg_C^+)^\# Rg_{(N-C)} \rightarrow (Rg_C^+)Rg_{(N-C-K)} + \kappa Rg \dots$$

where * denotes electronic excitation and # denotes vibrational excitation. Significant fragmentation is observed at ionizing energies greater than 100 meV above threshold. It should be noted that the evaporation times for the last few atoms may be quite long, so that experiments are often carried out on clusters with sufficient internal energy for further evaporation.

Conventional photoelectron spectroscopy of rare gas clusters involves the use of high energy photons, causing an electron to be ejected with non-zero kinetic energy. Measuring the kinetic energies of the photoelectrons gives information on the charged clusters, Rg_N^+, with the geometries of the neutral clusters (since, from the Born–Oppenheimer approximation, electron excitation and ejection is rapid on the timescale of atomic motions). As the electrons in the neutral cluster (held together by weak van der Waals forces) are localized on the atoms, the photoelectron is ejected (on a timescale of approximately 10^{-16} s) from a single atom.

The photoelectron spectrum of the icosahedral cluster Ar_{13}^+ has been found to be similar to those of larger argon clusters, as well as bulk argon. This has lead to the Ar_{13}^+ cluster being identified as an ionization chromophore for large clusters. This is not surprising, since the Ar^+–Ar interaction (Equation (2.8)) is proportional to r^{-4} and thus is small beyond the first shell of atoms around the ion.

2.3.4 Photoabsorption Spectra of Charged Rare Gas Clusters

Although neutral rare gas clusters are transparent in the visible region of the spectrum, charged rare gas clusters absorb strongly from the near infrared to the near ultraviolet regions. These absorptions are due to electronic transitions within the unit on which the positive charge is localized. The charged Rg_C^+ core can, therefore, be regarded as a discrete chromophore.

By measuring the photoabsorption spectra of Rg_N^+ clusters, one can determine the nature of the charged core, since Rg_2^+, Rg_3^+ and Rg_4^+ have very different optical spectra. The photoabsorption process for an $(N+C)$-atom charged rare gas cluster, with a C-atom core, can be written as:

$$(Rg_C^+)Rg_N + h\nu \rightarrow (Rg_C^+)^*Rg_N \rightarrow [(Rg_C^+)^*Rg_N]^\# \rightarrow (Rg_C^+)Rg_M + (N-M)Rg$$

Absorption of a photon leads to the chromophore (the cluster core) being electronically excited. This excited electronic state relaxes quickly, resulting in vibrational excitation of the cluster as a whole. This in turn leads to fragmentation—ejection of a certain number of rare gas atoms, which remove excess energy as kinetic energy, until a smaller, relaxed charge cluster is obtained.

For large argon clusters, it has been shown that the fragmentation process occurs by sequential evaporation of monomers (single rare gas atoms):

$$(Ar_N^+)^* \rightarrow (Ar_{N-1}^+)^* + Ar \rightarrow (Ar_{N-2}^+)^* + Ar \ldots$$

as observed for metal clusters. For smaller clusters however, single step fission, leading to the loss of two or more argon atoms, may also occur:

$$(Ar_4^+)^* \rightarrow Ar_2^+ + 2Ar$$

The measurement of the photoabsorption spectra of rare gas cluster ions involves selection of a specific ion, using a mass spectrometer. The selected clusters are irradiated by photons from a tuneable laser. When a spectral absorption occurs (i.e. when $h\nu$ corresponds to an optically allowed transition in the Rg_C^+ core), the cluster fragments, so the photoabsorption spectrum can be traced out by monitoring the frequency-dependent decrease in the original ion signal. Photoabsorption experiments by Haberland and co-workers have led to the identification of linear Xe_3^+ cores in Xe_{13}^+ and Xe_{19}^+ clusters (though a linear Xe_4^+ core has also been postulated for Xe_{19}^+).

2.3.5 Photofragmentation Spectra of Charged Rare Gas Clusters

As mentioned above, photoabsorption is generally accompanied by fragmentation of the cluster, via evaporation of neutral rare gas atoms. In a photofragmentation spectrum, a particular charged rare gas cluster is selected and irradiated by photons of fixed energy. All charged cluster species resulting from fragmentation are then detected by mass spectroscopy.

Figure 2.8 shows the photofragmentation spectrum of the Ar_{81}^+ ion upon irradiation with light of wavelength 610 nm (photon energy = 2 eV). At low photon flux (i.e. low laser power) only limited fragmentation occurs, and charged fragments are detected at around $N = 56\pm5$. Assuming single photon absorption at low photon flux, this means that absorption of 2 eV of energy results in the evaporation of approximately 25 atoms. Increasing the laser power results in photon absoption by the initially generated fragments, and further fragmentation to $N \approx 35\pm4$ atoms. Absorption of a third photon leads to fragments with $N \approx 18$. Experimental and theoretical studies have shown that the photofragmentation process is not statistical—i.e. there is a non-statistical distribution of energy in the electronic–vibrational and vibrational–kinetic energy transfer steps.

As the potential energy curves of the ground and excited states of the rare gas clusters do not cross, the cluster can only relax, vibrationally, as far as the minimum of the electronically excited state. After approximately 10^{-8} s (or longer if the excited state is metastable) the cluster emits a photon of lower energy than the excitation photon and further fragmentation may occur:

$$(Rg_M)^* \rightarrow Rg_{M-K} + K\,Rg + h\nu_2 \quad (\nu_2 < \nu_1)$$

The overall process:

$$Rg_N + h\nu_1 \rightarrow Rg_{M-K} + (N-M+K)Rg + h\nu_2$$

corresponds to fluoresence accompanied by fragmentation.

2.4.3 Localization of Excitation in Excited Rare Gas Clusters

As in the case of charged clusters, electronic excitation in rare gas clusters is generally localized on a linear core unit $(Rg_C)^*$—usually a dimer, trimer or tetramer, so the excited rare gas cluster $(Rg_N)^*$ can alternatively be written $(Rg_C)^*Rg_{N-2}$, with the excited core being solvated by ground state rare gas atoms. As in the excited argon dimer discussed above, the excited core can be regarded as a charged unit with a loosely associated electron—thereby constituting an electron-hole pair, or exciton in the language of solid state physics.

Jortner and colleagues have carried out experimental and theoretical studies of argon clusters containing xenon impurities (i.e. mixed clusters $XeAr_N$, with $N \approx$ 1,000). The fluorescence excitation spectrum of these clusters exhibits three broad bands which have been assigned to Xe atoms lying on top of the cluster surface, within the surface and in the interior of the cluster.

2.5 HELIUM CLUSTERS—LIQUID HELIUM DROPLETS

2.5.1 Liquid Helium Droplets

In recent years there has been much interest in the physics of helium clusters—which may also be described as liquid helium droplets, though the expression 'helium droplet' is normally reserved for helium clusters of 1000 or more atoms—and the spectroscopy of molecules dissolved in these clusters. This interest stems in part from the extensive research into bulk liquid helium and the fascinating topic of superfluidity.

Because of its weak vdW interactions and large zero-point energy (due to its low mass), quantum effects dominate the physics of helium at low temperatures and helium is the only substance that is known to remain liquid (at ambient pressure) right down to 0 K. In fact, helium can only be solidified at pressures above 25 atmospheres ($\approx 2.5 \times 10^6$ Pa). Helium is an ordinary, viscous liquid (He-I) just below its boiling point (4.2 K), but below a temperature of 2.18 K (for the ^{4}He isotope) or 3×10^{-3} K (for ^{3}He) a phase transition occurs to the superfluid (He-

II) state, which is characterized by vanishing flow resistance (i.e. zero viscosity), high heat conduction and quantized circulation. In the case of the ^{4}He isotope (a boson with nuclear spin $I = 0$), superfluidity is due to Bose condensation. For the ^{3}He isotope (a fermion with $I = \frac{1}{2}$), superfluidity may be due to the formation of quasi-Bose particles in the liquid.

There is considerable interest in determining whether finite liquid helium droplets (i.e. He clusters) should also exhibit superfluidity. Droplets of ^{4}He were first observed by Kamerlingh Onnes, as far back as 1908, during his attempt to liquify ^{4}He. In 1961, Becker and co-workers used molecular beam techniques to generate ^{4}He droplets, while in 1977 Gspann and co-workers produced a beam of ^{3}He droplets. In both cases, the He droplets were produced by free jet expansion of gaseous He through a nozzle (5–20 μm diameter), with a stagnation source temperature of 5–30 K and pressure of 5–$80{\times}10^5$ Pa. Under these conditions, ^{4}He clusters are produced with a temperature of 0.38 K, while the lighter ^{3}He clusters are produced with a lower temperature of 0.15 K. Comparison with the bulk superfluid temperatures would lead to the prediction that ^{4}He clusters should be superfluid liquid droplets at 0.38 K, but that ^{3}He clusters will be normal liquid droplets at 0.15 K.

2.5.1.1 Stabilities of ^{3}He and ^{4}He Clusters

In Section 2.1.2, it was assumed that the binding energy of a rare gas dimer is equal to the potential well depth. For the lightest rare gas dimers, however, it is important to take account of the zero-point vibrational energy. Thus for the ^{4}He$_2$ dimer, (which has been detected experimentally and studied theoretically) the potential well depth is 0.9 meV (corresponding to $T_d = 11$ K), but the zero-point vibrational energy ($\frac{1}{2}h\nu$) is very high for these light atoms, resulting in a net binding energy (D_0) of only 1.31 mK! Although a ^{3}He atom is chemically equivalent to ^{4}He and the strength of vdW bonding is virtually identical in ^{3}He and ^{4}He clusters, the smaller mass of ^{3}He means that the zero-point energy of ^{3}He clusters is greater than that of ^{4}He clusters. Thus, the helium dimer isotopomers, ^{3}He$_2$ and ^{3}He^{4}He have higher zero-point energies than ^{4}He$_2$ and indeed are unbound ($D_0 > \varepsilon$). The weakness of the binding in the ^{4}He$_2$ dimer also shows up in the fact that, although the calculated equilibrium separation (r_e) is 2.97 Å, the mean internuclear separation $\langle r \rangle$ is calculated to be as large as 51.9 Å! Thus, the ^{4}He$_2$ dimer can best be regarded as a weakly bound, very floppy vdW molecule.

The helium trimer cluster ^{4}He$_3$ has been detected experimentally by the groups of Toennies and Saykally. The binding energy (corrected for zero-point vibrational energy) of the trimer has been estimated to lie in the range 106–294 mK—i.e. two orders of magnitude higher than the ^{4}He$_2$ dimer. Thus, the trimer is significantly more stable than would be expected by treating it as the sum of three dimer interactions—i.e. there is a large three-body cohesive contribution to the binding energy (although the nature of this interaction is not fully understood at present). The ^{4}He$_3$ trimer has been described as a 'Borromean' system, after the three interlinked rings (symbol of the Italian Borromeo family) which have the property that the removal of one ring leaves the remaining two rings unconnected. The increased strength of binding in the helium trimer is also reflected in the much smaller expectation value $\langle r \rangle = 9.01$ Å of the internuclear distances in $^4He_3$, compared with that for $^4He_2$ (51.9 Å).

Clusters of ^4He have been calculated to be stable for all sizes, with the binding energy per He atom rising smoothly from 1.3×10^{-3} K for $^4He_2$ to 7.2 K for bulk ^4He (this binding energy is reached for clusters with $N \geq 10^4$ atoms). ^3He_N clusters with $N < 29$ atoms are unstable (unbound) as the total zero-point energy exceeds the cluster dissociation well depth. For larger ^3He clusters, large oscillations are observed in the binding energy per atom—due to nuclear-spin pairing effects (as the ^3He nucleus is a fermion rather than a boson)—until convergence is reached on the bulk value (2.7 K). The lower binding energy of bulk liquid ^3He, compared with ^4He, is consistent with the lower temperature of generated ^3He clusters.

Calculations indicate that some manifestations of superfluidity occur for ^4He clusters with as few as 69 atoms. The gradual appearance of a roton mode for ^4He_N clusters with $N \gtrsim 100$ has been interpreted as experimental evidence for the onset of superfluidity. Calculations on mixed ^3He-^4He droplets indicate that spontaneous isotopic separation occurs, producing a droplet with a ^4He core surrounded by ^3He. This has also been observed experimentally.

2.5.1.2 Ionization of He Clusters

Most experiments on He clusters (droplets) involve mass spectroscopic detection of charged clusters generated by *electron impact ionization*. The cross section for ionization is proportional to the cross sectional area of the droplet. Initial He atom ionization is rapidly followed (within 1 ps) by the formation of a He_2^+ ionic core and the liberation of 2.2 eV, which has been shown to be enough to completely evaporate a droplet of approximately 4,000 atoms. The He_2^+ dimer dominates the mass spectra of ionic He clusters, with the intensities of larger charged clusters falling almost exponentially with increasing size. Superimposed on this trend, there are local maxima for $N = 2$, 7, 10 and 14 atoms—for both ^3He and ^4He clusters—indicating that these are magic numbers for He_N^+ clusters.

Photoionization measurements on liquid He droplets reveal an ion appearance potential of approximately 24.6 eV—virtually the same as the ionization potential of a free He atom (24.59 eV). Mass spectra after photoionization are very similar to those arising from electron impact ionization, indicating a similar fragmentation mechanism.

2.5.1.3 Excitation of He Clusters

Fluorescence excitation measurements on liquid He droplets show broad absorptions at 20.9 and 21.5 eV, which have been attributed to the atomic-like $2s$ and $2p$ states. Bands due to fluorescence from $(He_2)^*$ have also been observed. There is also evidence for the expulsion of the excited species He* and $(He_2)^*$ from the droplet.

2.5.2 Doped Liquid Helium Droplets

The first doping experiments on helium droplets were performed by Becker and co-workers in the 1980s. He clusters can be loaded with dopant atoms and molecules (D) by a pick-up experiment, where preformed He clusters are passed through a chamber containing vaporized dopant atoms or molecules, which are then adsorbed onto the cluster. Such experiments have enabled the study of metastable radicals, isomers and high spin clusters, such as Na_3 with three unpaired electrons.

As the strength of the $D \cdots He$ interaction is greater than the $He \cdots He$ interaction, the adsorption process is accompanied by the evaporation of many (often thousands) of He atoms:

$$He_N + D \rightarrow (D)(He_N)^* \rightarrow (D)He_M + (N-M)He$$

2.5.2.1 Location of Dopant Atoms and Molecules

Theory and experiment have shown that open shell dopant atoms (e.g. alkali metals) and molecules (e.g. O_2) lie on the surface of liquid helium droplets, due to strong repulsive interactions between the unpaired electrons and the He atoms. The surface of the droplet is deformed in the vicinity of the dopant—forming a dimple. Similar interactions (which are a consequence of quantum mechanical Pauli exchange repulsion) are responsible for the large (3.4 nm diameter) bubbles or voids found around electrons in bulk liquid helium. Electron bubble states have also been observed in the interior of large, finite liquid He droplets ($N > 5 \times 10^5$ atoms), though there have been predictions of stable surface states for electrons on such droplets. Smaller bubbles (with diameters of 0.5–0.6 nm) have been found for excited He atoms, metal atoms and alkaline earth metal ions in liquid He. (The smaller bubble diameters in these cases arise due to the balance of Pauli repulsion and attractive electrostatic and polarization forces between He and the ionic cores.)

Closed shell atoms and molecules (and most positive ions) are found at the centre of the He droplet. Cations have strong attractive interactions with neighbouring He atoms, leading to an increase of the He atom density relative to bulk He. In mixed $^3He/^4He$ clusters, therefore, dopant molecules such as SF_6 are observed to preferentially occupy the ^4He-rich core.

The neutral alkaline earth atoms (e.g. Ca, Sr and Ba) are found to lie close to (but not on) the surface of He droplets, though they have closed sub-shell (ns^2) electron configurations.

2.5.2.2 Spectroscopy of Dopant Molecules

The first spectroscopic investigation of doped He droplets was made by the Scoles group at Princeton in the early 1990s. Since this time, many experimental studies of the spectroscopy of molecules in helium droplets have been made by this group and by the group of Toennies at Göttingen.

As energy transfer from dopant molecules to the He droplet can be very rapid, evaporation of He atoms can also result in effective cooling of the adsorbed dopant molecule. Thus, liquid He droplets can be said to act as ideal matrices for performing spectroscopy on cold molecules.

Electronic, vibrational and rotational spectra have been measured for a number of molecules in liquid helium droplets. Direct absorption measurements are infeasible due to the low number density (10^{10}–10^{11} cm^{-3}) of doped droplets. Instead, photodepletion measurements are made. As mentioned in Chapter 1, photodepletion is a very sensitive technique. It relies on rapid transfer of photon-induced excitation from the absorbing molecule to the helium droplet. This leads to cluster heating and rapid ($\leq 10^{-6}$ s) evaporation of several hundred He atoms:

$$(\text{Mol})\text{He}_N + h\nu \rightarrow (\text{Mol})(\text{He}_N)^* \rightarrow (\text{Mol})(\text{He}_M) + (N-M)\text{He}$$

The spectrum is determined by measuring the decrease in intensity of the doped cluster peak as a function of the photon frequency.

Laser-induced fluorescence (LIF) measurements have also been made on doped He droplets. LIF gives a 10–100 times better signal-to-noise ratio in the visible region of the spectrum than photodepletion measurements and the results are more reliable for larger droplets ($N > 10^4$).

One of the most interesting results that have arisen from the spectroscopy of dopant molecules in liquid ^{4}He droplets, is the sharpness of spectral lines, with line widths as narrow as 100 MHz (0.03 cm^{-1}) being measured. Scoles and co-workers detected sharp, well-resolved rotational fine structure in the IR spectra of molecules such as SF$_6$ and OCS. Such fine structure is not observed in normal liquids and indicates that molecules can rotate freely in liquid ^{4}He droplets, due to their being superfluid. The same effect is observed in superfluid bulk liquid He. Analysis of the rotational lines reveals a decrease in the rotational constant of the molecule relative to the free (gas phase) value—corresponding to an increase (by two or three times) of the moment of inertia of the molecule in the He droplet. This has been attributed to the tight coordination of a number (e.g. eight in the case of SF$_6$, or two in the case of benzene) of He atoms to the molecule. However, the He atoms surrounding the molecule readily relax into a unique ground state, resulting in a completely homogeneous environment for the molecule, which explains why no splitting of spectral lines is observed in superfluid liquid He or He droplets.

The above results were observed in ^{4}He liquid droplets. However, under analogous conditions ^{3}He droplets are not superfluid, as their temperature (0.15 K) is significantly higher than the bulk superfluid temperature of liquid ^{3}He (0.003 K), which is manifest in broad peaks in the IR spectrum of, for example, OCS ($\delta\nu = 0.1$ cm^{-1}). Interestingly, it has been shown that the addition of sixty ^{4}He atoms to (OCS)^{3}He$_N$ results in a sharpening of the spectral lines and the reappearance of rotational fine structure. This has been interpreted in terms of the ^{4}He atoms lying at the core of the droplet and completely surrounding (solvating) the OCS molecule, as shown schematically in Figure 2.10. As the temperature of the cluster (presumably still around 0.15 K) is below the superfluid temperature of bulk liquid ^{4}He (2.18 K), the ^{4}He core of the droplet is superfluid, even though the ^{3}He mantle is not.

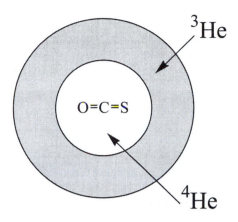

Figure 2.10 Schematic representation of an OCS molecule in a mixed liquid ^{3}He/^{4}He droplet.

2.5.2.3 Reactions in Helium Droplets

A number of groups have studied the use of liquid He droplets as media for controlling and manipulating reactions on a microscopic scale. Na atoms (which are found on the surface of the droplet) are mobile, as evidenced by the formation of Na$_2$ dimers. Cluster formation in He droplets is discussed in the following section.

The Toennies group have looked at the following exothermic reaction in ^{4}He droplets:

$$Ba + N_2O \rightarrow BaO + N_2$$

In order to generate cold products, Xe atoms are added to the He droplet before the reactants. As shown in Figure 2.11, the Xe atoms cluster at the centre of the droplet and attract the Ba atoms to the centre (due to strong vdW forces)—thereby overcoming the tendency of Ba to lie close to the surface of He droplets. The N$_2$O molecule, which is closed shell and, therefore, heliophilic, also penetrates the droplet—where it reacts on contact with the Ba atom to form a highly excited ($A^1\Sigma^+$) BaO molecule with approximately 2 eV of rovibrational energy. The BaO molecule is rapidly cooled by the evaporation of around 3,200 He atoms, leaving it in the $v' = 0$ state. The BaO molecule then radiates to the ground electronic state. In the absence of Xe, the reaction to form BaO occurs closer to the surface of the droplet, where less efficient cooling can take place, so hotter molecules are produced.

This experiment shows that He droplets may be used to isolate and study bimolecular reactions at very low temperatures (0.38 K) in superfluid liquid ^{4}He droplets. The added Xe, in this particular case, can be regarded as a catalyst of the reaction within the droplet, by helping to attract the reactants together.

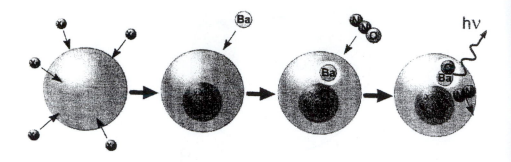

Figure 2.11 Representation of the reaction of Ba with N_2O in a Xe-doped liquid ^{4}He droplet.

2.5.2.4 Cluster Formation in Helium Droplets

Toennies and co-workers have made extensive use of liquid He droplets to form atomic and molecular clusters. The droplets pick up dopant species one at a time. Because of the low viscosity of (especially the superfluid ^{4}He clusters) He droplets, the dopants can move rapidly within the droplet and (within nanoseconds) can coagulate to form dopant clusters. The Toennies group have used this approach to generate clusters such as $(SF_6)_4$, $(H_2O)_{16}$ and Ag_N, as well as mixed clusters. The size of cluster formed is limited by the initial size of the He droplet, as dopant adsorption (and clustering) releases energy that leads to the evaporation of hundreds or thousands of He atoms. Large metal clusters of silver and europium (with masses up to 4,000 amu) have been produced by pick-up experiments on liquid He droplets and this opens up the possibility of studying metal clusters which are generated at very low temperatures.

2.6 EXERCISES

2.1   Explain why charged and electronically excited rare gas dimers have approximately equal dissociation energies and equilibrium interatomic distances.

2.2   If the zero-point energy of the ^{4}He$_2$ dimer is 1 (in arbitrary units), calculate the zero-point energies of the ^{3}He^{4}He and ^{3}He$_2$ dimers. (Hint: assume that the He$\cdots$He interaction is harmonic and that the He$\cdots$He stretching force constant is the same for all three isotopomers.)

2.7 FURTHER READING

Haberland, H., 1994, Rare gas clusters. In *Clusters of Atoms and Molecules*, Vol.I, edited by Haberland, H., (Berlin: Springer), pp. 374–395.

Nelson, D.R. and Spaepen, F., 1989, Polytetrahedral order in condensed matter. *Solid State Physics*, **42**, pp. 1–90.

Phillips, J.C., 1986, Chemical bonding, kinetics, and the approach to equilibrium structures of simple metallic, molecular and network microclusters. *Chemical Reviews*, **86**, pp. 619–634.

Rigby, M., Smith, E.B., Wakeham, W.A. and Maitland, G.C., 1986, *The Forces Between Molecules*, (Oxford: Oxford University Press).

Stone, A.J., 1996, *The Theory of Intermolecular Forces* (Oxford: Oxford University Press).

Sugano, S., 1991, *Microcluster Physics*, (Berlin: Springer-Verlag), pp. 104–117.

Toennies, J.P. and Vilosov, A.F,. 1998, Spectroscopy of atoms and molecules in liquid helium. *Annual Review of Physical Chemistry*, **49**, pp. 1–41.

CHAPTER THREE

Molecular Clusters

3.1 INTRODUCTION

3.1.1 What are Molecular Clusters?

In this chapter, the term 'molecular cluster' will be used to represent a cluster which is an aggregate of an integer number of molecules. The molecules may be all of the same type (homo-molecular clusters, such as the water clusters $(H_2O)_N$) or of different types (hetero-molecular clusters, such as $(CH_3OH)_M(H_2O)_M$). The term 'molecular cluster' will also be used to describe clusters composed of molecules and rare gas or other atoms or ions, such as $(Rg)_M(C_6H_6)_M$ and $(I^-)(H_2O)_N$. Molecular clusters should not be confused with cluster molecules, which are the covalently bonded clusters and cages (such as boron hydride and transition metal carbonyl clusters) of inorganic chemistry and which are outside the scope of this book.

3.1.2 Why Study Molecular Clusters?

Molecular clusters are of interest for a number of reasons. First as models for important processes (such as ozone depletion, acid rain formation and the generation of pollutants) occurring in the atmosphere, many of which take place on the surface of water droplets. Second as models for solvation, where solvent effects on the electronically and vibrationally excited states of a particular solute molecule (containing an appropriate chromophore group) may be investigated as a function of the the nature and number of solvent molecules in the cluster. Related questions concern the localization and transfer of charge and excitation in molecular clusters. In contrast to atomic clusters (such as those of the rare gases or metallic elements), fragmentation patterns accompanying ionization or excitation of molecular clusters may be modified by rapid chemical reactions between the ion/excited molecule and its surrounding molecules. The study of the geometries and stabilities of molecular clusters, as a function of size, may give important information on nucleation and crystal growth phenomena.

3.1.3 Intermolecular Forces

Most molecular clusters are composed of molecules which are stable and which possess closed electronic shells—notable exceptions being the open shell molecules NO (with one unpaired electron and a $^2\Pi$ ground state) and O_2 (with two unpaired electrons and a $^3\Sigma$ ground state). There are two distinct bonding modes in a molecular cluster: within each constituent molecule there are strong covalent bonds; but the cluster is held together by much weaker intermolecular forces.

Types of intermolecular forces, and examples of clusters held together by these forces, are given below. Some interactions, such as dispersion forces, are also found in atomic clusters, while others, such as multipolar induction forces and hydrogen bonding, are unique to molecular clusters.

3.1.3.1 Dispersion Interactions

One component of the intermolecular bonding in all molecular clusters is the dispersion energy, arising due the interaction of instantaneous atom-centred dipoles, as discussed in Chapter 2 for rare gas clusters. In the case of clusters composed of homonuclear diatomic and polyatomic molecules, such as $(I_2)_N$ and $(S_8)_N$, these dispersion forces dominate the intermolecular bonding. As molecules are larger and (generally) more polarizable than rare gas atoms, typical binding energies due to dispersion forces are of the order $\varepsilon \lesssim 100$ meV ($\lesssim 10$ kJ mol^{-1}) per intermolecular interaction—i.e. $\varepsilon(Mol_N) \approx 10\varepsilon(Rg_N)$. For example, the magnitude of the dispersion energy for two methane molecules separated by 0.3 nm is approximately -4.7 kJ mol^{-1} (i.e. $\varepsilon \approx 4.7$ kJ mol^{-1}).

3.1.3.2 Dipole–Dipole Interactions

A polar molecule is one which possesses a permanent dipole moment, that is, there is an asymmetric charge distribution, with one end of the molecule relatively negative ($-q$) with respect to the other ($+q$)—examples being the diatomic molecule F–Cl (where the F atom is negative with respect to the Cl atom) and the polyatomic molecule HCCl$_3$ (where the H end of the molecule is positive with respect to the three Cl atoms). The magnitude of the dipole moment (which by definition points from the negative toward the positive end) is given by $\mu = q\ell$, where ℓ is the separation between the centres of the two opposite charges. When two polar molecules are close to each other there is a tendency for their dipoles to align, with the attractive head-to-tail arrangement being lower in energy than the repulsive head-to-head or tail-to-tail arrangements, as shown in Figure 3.1a. The energy of interaction of two co-linear dipoles (μ_1 and μ_2), arranged in a head-to-tail fashion, is given by:

$$V^{DD}(r) = -\frac{2\mu_1\mu_2}{4\pi\varepsilon_0 r^3} \tag{3.1}$$

where r is the distance between the centres of the dipoles. The potential energy of interaction between two parallel dipoles which are not co-linear (i.e. where the inter-molecular vector makes an angle θ to the first intra-molecular bond) is given by:

$$V^{DD}(r) = -\frac{2\mu_1\mu_2}{4\pi\varepsilon_0 r^3}\left(1 - 3\cos^2\theta\right) \tag{3.2}$$

For general intermolecular geometries, a function of the three angles that define the relative orientations of the two dipoles, must be included in this equation. In the side-on arrangement, antiparallel dipoles are favoured energetically over parallel dipoles.

The alignment of molecular dipoles is opposed by the effect of thermal motion. At room temperature, where the thermal energy (kT) is greater than the electrostatic dipole–dipole intrreaction energy, the average energy of interaction between two dipoles μ_1 and μ_2, seperated by a distance r, can be obtained from a Boltzmann-weighted average:

$$V^{DD}(r) = -2\left(\frac{\mu_1\mu_2}{4\pi\varepsilon_0}\right)^2 \left(\frac{1}{r^6}\right)\left(\frac{1}{3kT}\right) \tag{3.3}$$

At room temperature (T = 300 K), Equation (3.3) becomes:

$$V^{DD} / \text{kJ mol}^{-1} = -\frac{(\mu_1 / D)^2 (\mu_2 / D)^2}{1025(r / \text{nm})^6} \tag{3.4}$$

where the dipole moments are in Debye (1 D $=$ 3.336×10^{-30} C m). For two molecules with dipole moments of 1 D (e.g. HCl), at a separation of $r = 0.3$ nm, this gives an average dipole–dipole interaction energy of approximately −1.4 kJ mol^{-1}. It should be noted, however, that clusters of dipolar molecules are usually generated at low temperatures, where there is no thermal averaging, so the interaction energies will be somewhat higher.

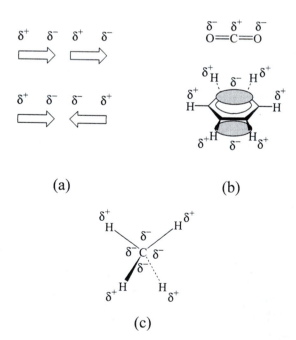

(a)   (b)

(c)

Figure 3.1 Molecular multipoles. (a) Attractive and repulsive interactions between molecular dipoles. (b) Examples of molecular quadrupoles. (c) The molecular octopole of CH$_4$.

As well as dipole moments, molecules may possess higher order multipole moments, arising from their non-spherical charge distributions. Thus, although the linear molecules CO_2 (O=C=O) and acetylene (H–C≡C–H) and the planar molecule benzene (C_6H_6) do not have dipole moments, they have non-zero quadrupole moments, as shown in Figure 3.1b. For more symmetrical molecules, the first non-zero multipole moments have higher order: thus, the methane molecule (CH_4) has no dipole or quadrupole moment, but it has a non-zero octopole moment (Figure 3.1c). The electrostatic interaction between higher order multipoles can be treated in an analogous fashion to that described above for dipole–dipole interactions.

Although not necessarily dominating the binding energy, high-order multipole interactions may play an important role in determining the structure of molecular dimers and higher clusters. Thus, a study of the $(HF)_2$ dimer by Dyke and co-workers, using microwave spectroscopy, revealed that the dimer does not have the linear H–F···H–F configuration, which would be predicted on the basis of exclusively dipole–dipole interactions. In the experimentally determined structure (see Figure 3.2), the bonding with respect to one of the HF molecules is approximately linear ($\theta_A = \angle$F–H···F ≈ 0), but the second angle ($\theta_B = \angle$H···F–H) is close to 60°. The observed structure is stabilized by the combination of dipole–dipole, dipole–quadrupole and quadrupole–quadrupole interactions.

Figure 3.2 The experimentally determined geometry of the $(HF)_2$ dimer.

In cases where quadrupolar interactions dominate, T-shaped intermolecular geometries are generally adopted, with the positive regions of one quadrupole being attracted to the negative regions of another. Such a situation applies to the benzene dimer $(C_6H_6)_2$, which has a T-shaped geometry where one C–H bond of one molecule is oriented towards the π-electron cloud of the other, as shown in Figure 3.3a. (In the benzene molecule, the ring C atoms are relatively negative with respect to the H atoms.) The geometry of the benzene–benzene interaction has, however, also been attributed to a novel C–H···π hydrogen bond. Interestingly, the quadrupole in perfluorobenzene (C_6F_6) is the opposite way round to that of benzene (i.e. the peripheral F atoms carry more electron density than the C atoms of the ring). As expected, mixed C_6H_6/C_6F_6 solids have alternating π-stacks of C_6H_6 and C_6F_6 rings, an arrangement which is preferred due to the reversed quadrupoles. Similar interactions are expected for mixed $(C_6H_6)(C_6F_6)$ dimers and clusters, as shown in Figure 3.3b.

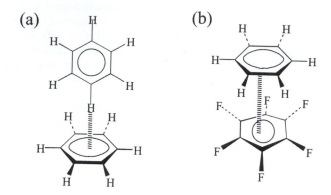

Figure 3.3 Benzene dimers. (a) T-shape interaction in $(C_6H_6)_2$. (b) π-stacking interaction predicted for the mixed dimer $(C_6H_6)(C_6F_6)$.

3.1.3.3 Induction Forces

The assymmetric charge distribution of a polar molecule leads to the polarization of neighbouring molecules (which may themselves be polar or non-polar) and the creation of an induced dipole, $\mu = \alpha E$ (where α is the polarizability of the second molecule and E is the electric field due to the first, polar molecule). The induced dipole interacts in an attractive fashion with the original dipole. As the direction of the induced dipole follows that of the permanent dipole, the effect survives even if the permanent dipole is subject to thermal reorientations, so no temperature averaging occurs. The interaction energy between a permanent dipole (μ_1) on molecule 1 and an induced dipole on molecule 2 (which has polarizability α_2), at a distance r, is given by:

$$V^{\mathrm{DID}}(r) = -\frac{2\mu_1^2\alpha_2}{\left(4\pi\varepsilon_0\right)^2 r^6} = -\frac{\mu_1^2\alpha_2'}{2\pi\varepsilon_0 r^6} \tag{3.5}$$

where α_2' is the polarizability volume of molecule 2 ($= \alpha_2/(4\pi\varepsilon_0)$). Equation (3.5) can be rewritten in the form:

$$V^{\mathrm{DID}}/\mathrm{kJ\,mol^{-1}} \approx -\frac{\left(\mu_1/\mathrm{D}\right)^2\left(\alpha_2'/10^{-24}\ \mathrm{cm}^3\right)}{1.66\times10^4\left(r/\mathrm{nm}\right)^6} \tag{3.6}$$

For a molecule with $\mu \approx 1$ D (e.g. HCl), neighbouring a molecule with polarizability volume $\alpha' \approx 10\times10^{-24}$ cm^3 (e.g. benzene), the dipole–induced–dipole interaction energy is approximately -0.8 kJ mol^{-1} at $r = 0.3$ nm.

Dipoles (and higher multipoles) can also be induced by neighbouring molecules with permanent charges (monopoles) or high-order multipoles, and similar expressions can be derived for the interactions between these permanent and induced multipoles.

3.1.3.4 Hydrogen Bonding

A hydrogen bond is an intermolecular linkage of the form X–H$\cdots Y$, where a hydrogen atom is covalently bound to one electronegative atom (X = N, O, F etc.) and interacts with a second electronegative atom (Y), which generally has an accessible lone-pair of electrons.

Characteristic experimental properties of hydrogen bonds are: the $\angle X$–H$\cdots Y$ angle is normally close to 180°; the H$\cdots Y$ distance is significantly longer than a H–Y covalent bond; lengthening of the X–H bond; a large red-shift of the X–H stretching vibration; broadening and gain of intensity of the X–H stretch; and deshielding of the hydrogen-bonded proton (observed as a shift in its NMR signal).

There are a number of contributions to hydrogen bonding. The major component (probably around 80%) is electrostatic, as a H atom bonded to an electronegative atom (X), will carry a significant fractional positive charge (δ^+) which will undergo an attractive interaction with the fractional negative charge (δ^-) on the neighbouring electronegative atom (Y). Other important contributions come from induction and dispersion forces, as well as from charge transfer— which can be regarded as incipient covalent bond formation—from atom Y to the hydrogen atom. Typical hydrogen bond strengths lie in the range 10–25 kJ mol^{-1}.

Hydrogen bonding has very important consequences for the structures and energetics of the hydrides of light elements. The anomalously high boiling points of ammonia, water and hydrogen fluoride can be attributed to the occurrence of hydrogen bonding, which is significantly stronger for these elements than for their heavier congeners. The tetrahedral three-dimensional structure of ice and the local structure of liquid water is a consequence of hydrogen bonding, as is the fact that ice is less dense than water! The very existence of life on Earth is a direct consequence of the hydrogen bonding present in water and the high range (100°C) over which water remains liquid. In fact, hydrogen bonding plays a very important rôle in biology, being responsible for the structural organization of proteins and other bio-molecules and governing crucial *in vivo* proton-transfer reactions. Hydrogen bonding makes a significant contribution to the binding of clusters of protic molecules such as (HF)$_N$, (H$_2$O)$_N$, (NH$_3$)$_N$ and (ROH)$_N$, giving rise to typical cluster binding energies ε of up to 250 meV (25 kJ mol^{-1}) per hydrogen bond.

3.1.3.5 Comparison of Intermolecular Forces

In many molecular clusters, more than one of the types of intermolecular force described above may be in operation, depending upon whether the molecules carry a net charge or possess permanent multipole moments. With the exception of small, highly polar molecules (such as H$_2$O), the dispersion energy is the largest contribution to intermolecular bonding in molecular clusters. Induction energies, on the other hand, are small unless one or more of the species in the cluster are charged. The relative strengths of the intermolecular forces found in molecular clusters can be appreciated from Table 3.1, which shows effective Lennard–Jones potential well depths (expressed as the equivalent temperature ε/k) for a variety of molecules, and compares them with the rare gases neon, argon and xenon.

Table 3.1 shows that small, non-polar molecules (such as H_2 and N_2) have low intermolecular binding energies (weaker than the interatomic forces in argon, for example), while larger molecules (e.g. CCl_4 and C_6H_6) have significantly higher interactions. The table also shows that the boiling points of these molecular sytems follow the same general trend as the well depth. Hydrogen bonded dimers have significantly higher well depths—for example an estimate of the effective pair potential well depth for $H_2O \cdots H_2O$ gives a value of 2,400 K, which is consistent with the relatively high boiling temperature (373 K) of liquid water.

Table 3.1 Comparison of boiling points and effective Lennard–Jones potential well depths for atomic and molecular dimers.

	(ε/k) / K	T_b / K		(ε/k) / K	T_b / K
Ne	36	27	CO_2	190	195*
Ar	124	87	CH_4	137	112
Xe	229	166	CCl_4	327	350
H_2	33	20	C_6H_6	440	353
N_2	92	77	H_2O	2400	373

*CO_2 sublimes at atmospheric pressure.

3.2 EXPERIMENTAL INVESTIGATION OF MOLECULAR CLUSTERS

Before turning to a discussion of specific types of molecular cluster, I will introduce some general features, relevant to the experimental investigation of molecular clusters.

3.2.1 Ionization of Molecular Clusters

The ionization of molecular clusters proceeds via the ionization of a single molecule in the cluster, a process which is generally accompanied by molecular (electronic and/or vibrational) excitation:

$$(Mol)_N \rightarrow [(Mol)_N^+]^* \equiv (Mol)_{N-1}(Mol^+)^* + e^-$$

Relaxation of the initially generated ion is accompanied by substantial intramolecular and intermolecular configurational changes. Relaxation mechanisms include cluster fragmentation—i.e. the loss of intact molecules from the cluster:

$$[(Mol)_N^+]^* \rightarrow (Mol)_{N-1}^+ + Mol$$

and dissociative ionization, which involves fragmentation of molecules within the cluster, accompanied by loss of one or more fragments. Common examples of this type of reaction include proton transfer in hydrogen bonded clusters of the form $(HB)_N^+$, where $B = $ OH, NH_2, etc.:

$$[(HB)_N^+]^* \equiv (HB)_{N-1}(HB^+)^* \rightarrow (HB)_{N-1}H^+ + B$$

though transfer of other species than protons is possible:

$$\text{e.g.} \quad [(CS_2)_2^+]^* \equiv (CS_2)(CS_2^+)^* \quad \rightarrow CS_3^+ + CS$$
$$\rightarrow S_2^+ + 2CS$$
$$\rightarrow C_2S_3^+ + S$$
$$\rightarrow CS_4^+ + C$$

Ionization of clusters consisting of a large aromatic molecule embedded in a solvent cluster, where the solvent may be a rare gas, is accompanied by relatively small intra- and intermolecular configurational relaxation. This is because there is delocalization of the positive charge in the aromatic molecule, leading to a small change in the bonding between the aromatic molecule and the solvent atoms/molecules on ionization—i.e. the induction forces are small.

3.2.2 Spectroscopy of Molecular Clusters

Molecular clusters contain obvious chromophores (the constituent molecules) which act as localized centres for absorbing radiation. This enables the study of electronic and vibrational excitation of molecules in a cluster environment—either by photodepletion experiments or by direct absorption spectroscopy.

3.2.2.1 Vibrational Spectroscopy of Molecular Clusters

One fundamental property of molecular clusters is that the intramolecular (covalent) bonds are much stronger (for example, typical dissociation energies for diatomic molecules lie in the range 1–3 eV, or 100–300 kJ mol^{-1}) than the intermolecular van der Waals binding energies (which are generally less than 300 meV or 30 kJ mol^{-1}). Thus, in many cases, vibrational excitation of a molecule (in a molecular cluster) by a single quantum will impart sufficient energy to cause partial photodissociation of the cluster—i.e. $h\nu_{vib}(Mol) > E_b(Mol_N)$:

$$(Mol)_N + h\nu \rightarrow [(Mol)_N]^{\#} \rightarrow (Mol)_{N-M}^+ + M(Mol)$$

In cases where direct photoabsorption spectra cannot be measured, IR spectra can be obtained by a photodepletion experiment. Neutral molecular clusters are mass selected by collision with a beam of atomic helium and the spectrum is mapped out by scanning the frequency of an IR laser and monitoring the decrease in MS intensity as a function of the laser frequency.

Quantities which can be extracted from these experiments include:

- *line shifts*—due to interactions between the excited (solute) molecule and its neighbouring (solvent) molecules—this provides information on the cluster structure;

- *line widths*—due to dissociation of the clusters—this provides information on the lifetimes of excited states and, hence, on the dynamic coupling of intramolecular and intermolecular vibrational modes;

- *line intensities*—allow photodissociation cross sections to be obtained, which are directly related to the mechanisms of the absorption and decay processes.

3.2.2.2 Electronic Spectroscopy of Molecular Clusters

Optical laser-induced fluorescence is a powerful technique for measuring the electronic spectra of molecules. As the technique is not size specific, however, in order to study electronic transitions in molecular clusters, two-colour resonant two-photon ionization (2CR2PI), combined with ion detection, is adopted instead. In this technique, a tunable laser (variable frequency v_1). Is used to electronically excite the molecular cluster. When the frequency corresponds to a resonance, the excited state lives long enough for the absorption of a second (fixed frequency v_2) photon, which leads to ionization (with low fragmentation). The ionized cluster is then detected by a mass spectrometer. In this way, the electronic spectrum of the neutral molecular cluster is mapped out. The overall process can be written:

$$(\text{Mol})_N + hv_1 \rightarrow [(\text{Mol})_N]^{\#} + hv_2 \rightarrow (\text{Mol})_N^+ + e^-$$

Such experiments have been used to investigate solvent effects on molecular electronic transitions in molecular clusters.

3.2.3 Electric Dipole Measurments

A number of hydrogen bonded clusters have been studied using electrostatic multiple focusing techniques. The clusters $(H_2O)_N$, $(HF)_N$ and $(NH_3)_N$ (with $N \geq$ 3) show negligible refocusing, indicating that they have undetectable dipole moments. This is consistent with non-polar cyclic, rather than polar linear structures. These experiments have also shown that the acetic acid trimer $(CH_3CO_2H)_3$ is polar, while $(CH_3CO_2H)_4$ is non-polar.

3.3 WATER CLUSTERS

The earliest studies of water clusters date back to the early 1950s, when Pimental measured the IR spectroscopy of matrix-isolated water clusters. In 1974, Dyke and colleagues measured the microwave spectra of small water clusters generated in a molecular beam. In recent years, however, the number of experimental and theoretical studies of water clusters has increased dramatically.

3.3.1 Mass Spectroscopic Studies of Water Clusters

Water clusters can readily be generated by evaporation, condensation and supersonic expansion into a molecular beam. Mass spectrometric measurements can then be made by electron impact or photoionization of the neutral water clusters.

A typical electron-impact mass spectrum of water clusters is shown in Figure 3.4a. The cationic species which are detected by the mass spectrometer are the protonated clusters $(H_2O)_N H^+$. By analogy with the rare gas clusters, discussed in Chapter 2, magic numbers can be detected—corresponding to drops in MS intensity after $N = 21$ and (to a lesser extent) after $N = 28$ and 30.

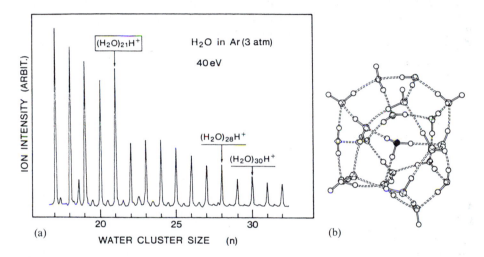

(a)   (b)

Figure 3.4 Protonated water clusters. (a) Electron impact (40 eV) mass spectrum of protonated water clusters $(H_2O)_N H^+$, in the range $N = 17$–32. Magic number clusters are indicated by arrows. (b) The clathrate-like structure of $(H_3O^+)(H_2O)_{20}$.

On the basis of infrared spectroscopic measurements, these protonated clusters have been shown to consist of a number of water molecules surrounding a hydronium ion, H_3O^+. Thus, the cluster $(H_2O)_N H^+$ is better written as $(H_3O^+)(H_2O)_{N-1}$. The large peak in the MS for $N = 21$ has been explained in terms of a clathrate-like pentagonal dodecahedron of 20 water molecules, with the hydronium ion lying at the centre, forming hydrogen bonds to three of the water molecules, as shown in Figure 3.4b. The $(H_3O^+)(H_2O)_3$ moiety is the smallest unit where the inner hydronium ion is completely hydrogen bonded—i.e. fully solvated. The hydrated hydronium ion is the dominant ionic species in some regions of the atmosphere and such clusters are believed to be very important in chemical processes which occur both in the lower and the upper atmosphere. In the solid state, crystals are also known which contain the species $(H_5O_2)^+ \equiv$

$(H_2O)_2H^+$ (which has a symmetrical hydrogen bond: $H_2O\cdots H^+\cdots OH_2$) and $(H_9O_4)^+$ $\equiv (H_3O^+)(H_2O)_3$ (which has C_{3v} symmetry, with a central hydronium ion bound to three water molecules).

3.3.2 Infrared Spectroscopy of Cationic Water Clusters

Direct gas phase IR absorption spectra have been measured for small hydrated hydronium clusters $(H_3O^+)(H_2O)_N$ ($N = 1$–5), demonstrating that the $N = 1$ cluster $(H_5O_2)^+$, as in the solid state, has a symmetrical hydrogen bond, and that the first hydration shell of H_3O^+ is complete at $N = 3$.

Due to the localization of the positive or negative charge in clusters of the form $(H_3O^+)(H_2O)_N$ and $(OH^-)(H_2O)_N$, there is a contribution to the binding from ion-dipole and induction effects, so that the binding energies of ionic clusters are significantly stronger than in neutral clusters, being comparable to weak covalent bonds. Castleman and co-workers have tabulated the enthalpies of hydration of a number of cations and anions. For the first step in the hydration of the hydronium ion:

$$H_3O^+ + H_2O \rightarrow (H_3O^+)H_2O$$

the enthalpy change is $\Delta H_1 = -151$ kJ mol^{-1}. The enthalpy changes for successive addition of water molecules, however, decrease (i.e. there is a decrease in the average hydrogen bond energy as N increases) due to the delocalization of the positive charge. Comparison of the enthalpies for the Nth step in the solvation of H_3O^+:

$$(H_3O^+)(H_2O)_{N-1} + H_2O \rightarrow (H_3O^+)(H_2O)_N$$

($\Delta H_1 = -151$ kJ mol^{-1}, $\Delta H_2 = -93$ kJ mol^{-1}, $\Delta H_3 = -71$ kJ mol^{-1}, $\Delta H_4 = -64$ kJ mol^{-1}, $\Delta H_5 = -54$ kJ mol^{-1}, $\Delta H_6 = -49$ kJ mol^{-1}, $\Delta H_7 = -43$ kJ mol^{-1}) do not, however, reveal any obvious discontinuity corresponding to the completion of the first solvation of the charged core. (Other experiments indicate that the first solvation shell is complete at $N = 3$.) Thus, even second shell water molecules have high binding energies.

The charged core of a hydronium-water cluster gives rise to absorptions which are significantly stronger than those due to the surrounding neutral water molecules. The O–H stretching bands of the charged H_3O^+ core have slightly lower frequencies (by ≈ 100 cm^{-1}) than the neutral water molecule. Upon solvation, some of the stretching bands of the charged core are strongly red-shifted (being lowered by 1000 cm^{-1} or more) in small clusters, as the participation of the O–H bonds in hydrogen bonding, leads to their being weakened. As the number of neutral water molecules increases, however, the H_3O^+ stretching bands move to higher frequencies (i.e. they are blue-shifted) again, presumably because the strength of individual hydrogen bonds become weaker when the total number of hydrogen bonds is greater.

3.3.3 Negatively Charged Water Clusters

Although the negatively charged water molecule is thermodynamically unstable, because of its negative electron affinity, it is known that electrons can be trapped in localized states in liquid water, and in ice. Injection of low energy (< 1 eV) electrons into a beam of water clusters, which are seeded with a cold inert gas, results in electron trapping and the formation of small negatively charged clusters $(H_2O)_N^-$ and $Rg(H_2O)_N^-$ ($N \geq 2$). In the absence of the cooling rare gas, the water clusters are generated with more internal energy, which prevents the formation of the smaller negatively charged clusters (the adiabatic electron affinity of the water dimer, for example, is only 17 meV). The electron affinity of water clusters increases with increasing cluster size, so that larger water clusters lead to greater stabilization of the solvated electrons. By making the water clusters colder, the critical cluster size at which negative ions are first stabilized, is reduced. Theoretical calculations indicate that for clusters with up to 32 water molecules, the excess electron occupies a surface site, while for larger clusters the electron occupies an internal site.

Larger negatively charged clusters can be studied by attachment of low (approximately zero) energy electrons to preformed cold water clusters. Mass spectral studies of $(D_2O)_N^-$ clusters, using electrons with approximately zero kinetic energy, have shown that there is a sharp increase in intensity at $N = 12$, reaching a maximum at around $N = 20$.

Electron attachment is a resonant process, with a resonant width less than 0.5 eV, so that negatively charged water clusters are generated in an excited, metastable, state. Relaxation to the ground state can be accomplished via evaporation of water molecules:

$$[(H_2O)_N^-]^* \rightarrow (H_2O)_{N-1}^- + H_2O$$

Increasing the electron injection energy to around 7 eV results in partial molecular fragmentation and the formation of negatively charged clusters, $(OH^-)(H_2O)_N$, where the negative charge is localized on a hydroxide anion, which is solvated by the remaining water molecules. As discussed above, increasing the kinetic energy of the electrons further leads to electron impact ionization, molecular fragmentation and the formation of protonated clusters.

3.3.4 Structure and Spectroscopy of Neutral Water Clusters

3.3.4.1 Electron Diffraction Studies of Large Water Clusters

The electron diffraction patterns which have been measured for water clusters (with $\langle N \rangle$ in the range 1,500–2,000) are similar to those for the metastable, low pressure cubic phase of bulk ice. (Note: the most stable phase of bulk ice has a hexagonal structure.) Smaller clusters (with $N \lesssim 300$), however, generally have amorphous, or highly disordered structures—in agreement with simulation studies—consisting of disordered three-, four-, five- and six-membered hydrogen bonded rings of water molecules. The ice structure, by contrast, has exclusively six-membered rings. The smallest clusters studied (composed of around 20 water molecules) were assigned structures based on the dodecahedral $(H_2O)_{20}$ unit (as discussed above for protonated water clusters), with five-membered hydrogen bonded rings. Interestingly, calculations by Wales and colleagues (using empirical potentials), predict that the lowest energy structure for $(H_2O)_{20}$ is actually composed of three fused pentagonal prisms, as shown in Figure 3.5, rather than the dodecahedron. However, these calculations indicate, that it is possible for cooling water clusters to get trapped into structures which are not the lowest in energy.

These electron diffraction results are consistent with the IR spectra measured by Devlin and co-workers for water clusters of up to 10,000 molecules. The largest clusters give spectra which are very similar to that of crystalline ice, while smaller clusters (with $N \approx 100$) have spectra reminiscent of amorphous ice. Calculations have also shown that the $(H_2O)_{1000}$ cluster has an ordered (ice-like) interior and a disordered surface, with strained crystal-like structure in the sub-surface region. Smaller clusters (of hundreds of molecules) are predicted to have a strained ice structure, while clusters of less than 100 molecules (which are dominated by their surfaces) have structures similar to that of amorphous ice.

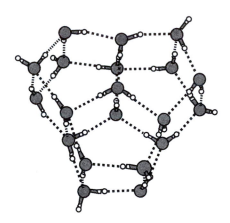

Figure 3.5 The fused pentagonal prismatic structure predicted by Wales and co-workers for $(H_2O)_{20}$.

3.3.4.2 Far-Infrared Spectroscopy of Small Water Clusters

Saykally and co-workers have carried out far-infrared vibration-rotation tunnelling (VRT) spectroscopy of small neutral water clusters $(H_2O)_N$ ($N = 3–5$). *Ab initio* MO calculations predict cyclic hydrogen bonded geometries for these clusters. The theoretically predicted equilibrium structures are shown in Figure 3.6, from which it can be seen that these low energy structures contain distinct proton environments—i.e. some protons are involved in hydrogen bonding, while others occupy terminal (dangling bond) positions.

According to theoretical calculations, there are at least two low energy mechanisms for interconverting the various protons in cyclic water clusters. The low energy barriers and the lightness of the hydrogen atoms, means that quantum mechanical tunnelling can occur through the potential barriers. This tunnelling leads to splitting of the vibration-rotation energy levels of the cluster. The absorption of far-IR radiation leads to transitions between these levels.

The lowest energy structure $(H_2O)_3$ for the trimer (and also the pentamer) is a frustrated structure. As shown in Figure 3.6, the odd-membered ring constrains there to be two dangling hydrogen atoms on adjacent atoms which are on the same side of the ring (leading to an unfavourable interaction between the parallel O–H dipoles). Such situations are avoided in even-membered rings, such as the tetramer, where the dangling hydrogen atoms can alternate above and below the molecular plane.

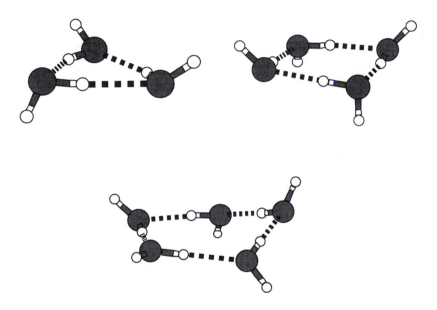

Figure 3.6 Lowest energy structures of $(H_2O)_3$–$(H_2O)_5$ which have been confirmed by far-IR VRT spectroscopy. Dashed lines represent hydrogen bonds.

The tunnelling splitting pattern of the water trimer has been rationalized by Wales in terms of two isomerization mechanisms. The lowest energy pathway (with an activation energy of 1 kJ mol^{-1}), which therefore leads to the larger tunnelling splitting (≈ 10 cm^{-1}), involves a flipping motion of a single water molecule—leading to interconversion of the terminal protons in the cluster structure. A second process involves rotation of a water molecule, and interconversion of hydrogen bonding and terminal protons, via a bifurcated transition state in which one molecule acts as a double hydrogen bond donor, while another acts as a double hydrogen bond acceptor. This second mechanism has a higher activation barrier (8 kJ mol^{-1}) and hence leads to a smaller tunnelling splitting ($\approx$ 0.01 cm^{-1}). The measured VRT spectrum, which is complicated, shows both splittings.

The tunnelling splitting patterns for the larger clusters are even more complex, and it is possible that other isomerization mechanisms may also contribute.

3.3.4.3 The Water Hexamer

Comparison of infrared spectroscopy and theoretical calculations of the $(H_2O)_6$ hexamer have led to the conclusion that the ground state structure has the 3-dimensional cage structure shown in Figure 3.7a. There are a large number of low-lying metastable isomers, such as the trigonal prismatic structure shown in Figure 3.7b which is almost isoenergetic with the cage structure, and the planar ring structure (Figure 3.7c) which is less stable by over 5 kJ mol^{-1}.

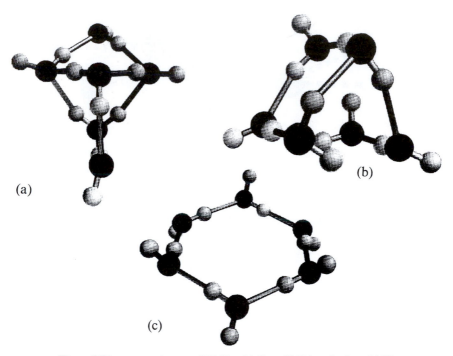

(a)

(b)

(c)

Figure 3.7 Low energy isomers of $(H_2O)_6$. (a) Cage. (b) Trigonal prism. (c) Ring.

On the basis of *ab initio* MO calculations, Clary and co-workers have shown that the intermolecular zero-point energy (ZPE) can be a very important quantity in molecular clusters, which are by definition quite weakly bound. The ZPE is defined as:

$$ZPE = \frac{1}{2}\sum_i h\nu_i \qquad\qquad (3.7)$$

where the summation is over all intermolecular vibration modes (with frequency v_i) of the cluster. In the case of the water hexamer, Clary has demonstrated that the ZPE is approximately one third of the cluster dissociation energy (D_e). Indeed, in many hydrogen bonded clusters, including some water clusters, zero-point energies may change the order of stability of isomers, compared to predictions based on D_e alone. The weakness of intermolecular interactions in molecular clusters also leads to the possibility of complex temperature dependence of isomer stability.

3.4 EXAMPLES OF OTHER MOLECULAR CLUSTERS

Although water clusters are probably the most widely studied clusters, many experimental studies have also been carried out on hydrogen-bonded ammonia and methanol clusters. Carbon dioxide clusters, on the other hand, serve as examples of non-hydrogen-bonded molecular clusters.

3.4.1 Ammonia Clusters

IR spectroscopy of protonated ammonia clusters $[(NH_3)_N H]^+$, has shown that there are strong absorption bands due to a central ammonium (NH_4^+) cation and weaker bands due to the remaining neutral ammonia molecules, so that the cluster should better be written: $(NH_4^+)(NH_3)_{N-1}$—with a central ammonium ion solvated by neutral ammonia molecules, as shown in Figure 3.8. This is analogous to the situation found for protonated water clusters. There is a strong red-shift of the N–H stretching modes of the NH_4^+ core, though the shifts of the vibrations of the outer NH_3 molecules are much smaller.

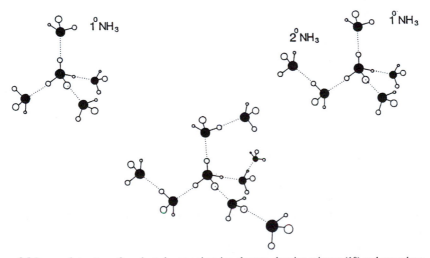

Figure 3.8 Proposed structures for solvated ammonium ion clusters, showing primary (1°) and secondary (2°) solvation shells.

The enthalpies of successive solvation of the ammonium cation:

$$(NH_4^+)(NH_3)_{N-1} + NH_3 \rightarrow (NH_4^+)(NH_3)_N$$

have been measured. By contrast with protonated water clusters, for protonated ammonia clusters, a discontinuity in ΔH_N is observed, between $N = 4$ and $N = 5$, corresponding to the completion of the first solvation shell at $N = 5$. The difference between water and ammonia clusters is presumably due to the greater strength of $O–H \cdots O$, compared with $N–H \cdots N$, hydrogen bonds.

Also in contrast to water clusters, negatively charged ammonia clusters, $(NH_3)_N^-$ are only observed for sizes ($N \geq 34$) for which internal electron solvation is possible, because surface electronic states are unstable in ammonia clusters. These solvated electron clusters of ammonia are finite analogues of the solvated electrons generated by dissolving sodium in liquid ammonia.

3.4.2 Methanol Clusters

Neutron diffraction studies of the α-phase of solid methanol have shown that it consists of infinite chains of hydrogen bonded CH_3OH molecules. Microwave spectroscopy measurements of small methanol clusters, however, reveal that methanol clusters with up to eleven molecules adopt non-planar ring structures, rather than chains. Beyond eleven molecules, the clusters have fused ring structures. The absence of three-dimensional cluster structures is because, unlike water or ammonia, methanol has only one protic hydrogen atom with which to form hydrogen bonds.

Theoretical simulations of liquid methanol predict that 73% of molecules take part in two hydrogen bonds—forming part of a chain or ring, while 19% form only one hydrogen bond (at the end of a chain) and 8% form three hydrogen bonds (where two rings are fused). Simulations of larger methanol clusters indicate that they are liquid-like, consisting of a mixture of rings and chains.

Photodepletion IR spectroscopy of small methanol clusters and analysis of the variation of the C–O stretching frequency as a function of cluster size, indicates that in $(CH_3OH)_N$ clusters with N in the range 3–5, the methanol molecules are symmetry equivalent, probably corresponding to planar hydrogen bonded rings. For $N \geq 6$, however, more than one stretching frequency is observed, perhaps reflecting reduced symmetry of these larger clusters.

3.4.3 Carbon Dioxide Clusters

The CO_2 molecule is linear and centrosymmetric and, therefore, possesses no dipole moment. Because the O atoms are relatively negatively charged with respect to the C atom, however, the molecule does possess a quadrupole moment (see Figure 3.1b). The asymmetric C=O stretching vibration provides a useful spectroscopic tag for investigating neutral and charged CO_2 clusters.

Lineberger and colleagues have studied the photodissociation of $(CO_2)_N^+$ clusters, where the chromophore is believed to be the $(CO_2)_2^+$ dimer. They found that the dissociation energy for the process:

$$(CO_2)^+_N + h\nu \rightarrow (CO_2)^+_{N-1} + CO_2$$

approaches a constant value of approximately 20 kJ mol^{-1} for $N \geq 13$. This should be compared with the sublimation enthalpy of approximately 24 kJ mol^{-1} for bulk CO_2.

Anionic $(CO_2)^-_N$ clusters have also been studied. Although the unsolvated CO^-_2 ion is unstable, it is stabilized by solvation by a neutral CO_2 molecule or by other molecules, such as water. Spectroscopic data indicates that the CO^-_2 ion in these complexes is bent, rather than linear (the geometry of CO_2).

3.5 SOLVENT–SOLUTE CLUSTERS

There has been much recent interest in studying the structures and spectroscopy of solvent–solute clusters, consisting of ions (either cations or anions) or neutral species, which are solvated by neutral molecules (usually water) or rare gas atoms. It is hoped that these studies will enable a better understanding of the fundamentals of structure, dynamics and reactions of ions and molecules in bulk solutions.

3.5.1 Solvated Ions

Self-solvation, as in protonated water clusters and hydroxide-water clusters, has been discussed above. Here, I will concentrate on clusters formed by ions of one species which are solvated by another. Such solvated cluster ions are of importance in the chemistry and physics of flames and combustion and also in the lower and upper atmospheres, as well as certain regions of space. As most experiments have been performed on hydrated ions, I will concentrate on these species. Many of the measurements of the thermodynamics, kinetics and reactivities of hydrated ions have been performed by Castleman and co-workers and his review (see References) should be consulted for further information.

Lisy and co-workers have formed solvated ion clusters by injecting ions (M^+ or X^-) into cold, preformed molecular clusters. Smaller metal cations give rise to greater polarization of the solvent molecules. In the case of $M^+(H_2O)_N$ clusters, smaller cations thus disrupt the structure of the water cluster to a greater extent, so that $H_2O \cdots H_2O$ hydrogen bonding is only observed for larger clusters. Similarly, in the clusters $(F^-)(H_2O)_N$ ($N = 3$–5), there is no evidence for hydrogen bonding. Larger halide anions (Cl^-, Br^-, I^-) are themselves polarized by the small, polar water molecules, leading to asymmetric solvation—i.e. the halide ions lie on the outside of the water cluster. Such external solvation is also favoured because it leads to less reorganization of the hydrogen-bonded water cluster. As the clusters get larger, eventually the charged species move to the interior of the cluster—i.e. they are internally solvated.

Table 3.2 Hydration enthalpies of alkali metal cations (kJ mol^{-1}).

	$\Delta_{hyd}H_1$	$\Delta_{hyd}H_4$	$\Delta_{hyd}H_{1-6}$	$\Delta_{hyd}H_b$
Li$^+$	-142	-69	-515	-520
Na$^+$	-100	-58	-403	-405
K$^+$	-75	-49	-333	-321
Rb$^+$	-67	-47	-306	-300
Cs$^+$	-57	-44	-282	-277

3.5.1.1 Hydration Enthalpies

The successive enthalpies of hydration of alkali metal cations (M^+; M = Li, Na, K, Rb and Cs) have been determined experimentally, by the Castleman group, by equating the hydration energy with the negative of the ion-water dissociation energy, obtained from high-pressure mass spectral studies of unimolecular dissociation. Some of their results are summarized in Table 3.2.

The Nth hydration of a cation corresponds to the following process:

$$M^+(H_2O)_{N-1} + H_2O \rightarrow M^+(H_2O)_{N-1} \qquad\qquad \Delta H = \Delta_{hyd}H_N(M^+)$$

Each successive hydration is less exothermic than those preceding it (e.g. $\Delta_{hyd}H_1(Li^+) = -142$ kJ mol^{-1}, $\Delta_{hyd}H_4(Li^+) = -69$ kJ mol^{-1}). There are no obvious breaks in hydration enthalpy as a function of N due to water shell formation. As the ion-water clusters are primarily held together by ion-dipole and induction forces, the hydration energies decrease with increasing ionic radius (e.g. $\Delta_{hyd}H_1(Li^+) = -142$ kJ mol^{-1}, $\Delta_{hyd}H_1(Na^+) = -100$ kJ mol^{-1}, $\Delta_{hyd}H_1(Cs^+) = -57$ kJ mol^{-1})—in agreement with the Born equation for the Gibbs free energy of solvation:

$$\Delta_{solv} G = -\frac{1}{2}\left(\frac{q_i^2 e^2}{4\pi\varepsilon_0 r_i} \right)\left(1 - \frac{1}{\varepsilon_r} \right) \qquad\qquad (3.8)$$

where q_i and r_i are the charge and radius of the ion respectively and ε_r is the dielectric constant (relative permittivity) of the solvent. Assuming the entropy of hydration is fairly constant for these cluster species, trends in $\Delta_{hyd}G$ will be mirrored by $\Delta_{hyd}H$.

The sum of the first six hydration energies (corresponding to the overall process):

$$M^+ + 6H_2O \rightarrow (M^+)(H_2O)_6 \qquad\qquad \Delta H = \Delta_{hyd}H_{1-6}(M^+)$$

is found to be very close to the hydration energies of the ions in bulk water (e.g. $\Delta_{hyd}H_{1-6}(Li^+) = -515$ kJ mol^{-1}; $\Delta_{hyd}H_b$ (Li$^+$) $= -520$ kJ mol^{-1}) and again follows the trend predicted by the Born equation. This would seem to indicate that the alkali metal cations are hexa-coordinate in bulk aqueous solutions, though this is in contradiction to coordination numbers obtained from transport measurements.

Similar experiments have been carried out to obtain the hydration enthalpies of the halide ions $X^- = F^-$, Cl$^-$, Br$^-$ and I$^-$ (see Table 3.3).

Table 3.3 Hydration enthalpies of halide anions (kJ mol^{-1}).

	$\Delta_{hyd}H_1$	$\Delta_{hyd}H_4$	$\Delta_{hyd}H_{1-6}$	$\Delta_{hyd}H_b$
F$^-$	-97	-56	-389	-506
Cl$^-$	-55	-46	-287	-364
Br$^-$	-53	-46	-280	-337
I$^-$	-43	—	-228	-296

As for the alkali metals, the hydration energies decrease with increasing ionic radius and successive hydration energies get smaller. It is noticeable, however, that the sum of the first six hydration enthalpies is significantly smaller than the bulk hydration enthalpy for these ions, perhaps indicating that more than 6 water molecules are intimately involved in the hydration of these anions in bulk aqueous solutions.

Comparing cations and anions of elements in the same period (row) of the periodic table, the anions have larger ionic radii (e.g. $r(F^-) > r(Li^+)$; $r(Cl^-) > r(Na^+)$ etc.) and, thus, from the Born equation, are expected to form weaker bonds to polar molecules. This is reflected in the smaller hydration energies of the halides compared with the alkali metal cations.

3.5.1.2 Non-aqueous Solvation

Castleman and colleagues have also measured the solvation energies of cations and anions in clusters of solvent molecules other than water. The measured binding energies for a variety of (Mol)Na$^+$ complexes follow the order DME > NH$_3$ > H$_2$O > SO$_2$ > CO$_2$ ≥ HCl > N$_2$ ≥ CH$_4$. DME is dimethoxyethane, which acts as a bidentate ligand, forming two O$\cdots$Na$^+$ interactions per molecule. Hereby explaining the high binding energy. The stronger binding energy of NH$_3$ to Na$^+$, compared with H$_2$O, is due to the larger ion-induced dipole in ammonia. SO$_2$ has a dipole moment roughly equal to that of NH$_3$, but the quadrupole interaction with cations is repulsive, leading to a weaker overall interaction. The relative strengths of binding of the non-polar molecules CO$_2$, N$_2$ and CH$_4$ to Na$^+$ reflect their decreasing polarizabilities.

As mentioned above, the larger radius of Cl$^-$, compared to Na$^+$, generally leads to lower interaction energies with polar solvent molecules. For SO$_2$ and HCl, however, the quadrupolar Mol$\cdots$Cl$^-$ interactions are attractive, rather than repulsive, so overall these molecules bind more strongly to Cl$^-$ than to Na$^+$.

3.5.1.3 Solvation of Ionic 'Molecules'

Another area of interest is to determine how many water molecules are required to stabilize charge separation of two oppositely charged ions. It has been determined that for NaCl, the smallest number of water molecules that can stabilize charge separation is six—Na$^+$(H$_2$O)$_6$Cl$^-$, though at this size the contact ion pair, (Na$^+$Cl$^-$)(H$_2$O)$_6$, is slightly lower in energy. *Microbrine* particles—saturated aqueous NaCl clusters—have approximately 54% ion pairing near the surface, where there is a slight excess of Cl$^-$ ions. In the sub-surface region, there is a corresponding excess of Na^+ ions, while the bulk interior has equal concentrations of Na^+ and Cl^- ions and approximately 75% ion pairing.

3.5.2 Solvated Neutral Atoms

There has been much interest in determining the properties of water and ammonia clusters doped by neutral sodium atoms. For clusters $Na(H_2O)_N$, the smallest clusters ($N \leq 4$) have ionization potentials which decrease with increasing N. The lowest energy ionization occurs from localized Na atoms. For larger clusters, however, ($N = 5$–20) ionization is only weakly size-dependent and occurs from surface-localized $Na^+{:}e^-$ ion pairs. For the analogous ammonia clusters, $Na(NH_3)_N$, the transition from Na-localized to ion-pair ionization occurs at around 10 ammonia molecules. These experiments indicate that, while 5 water molecules are sufficient to dissolve (and ionize) sodium atoms, 10 ammonia molecules are required to achieve the same dissolution. This is presumably linked to the reduced thermal stability of negatively charged ammonia clusters compared to those of water.

3.5.3 Solvated Molecules

There have been a number of studies of the dissociation dynamics of diatomic molecules, such as I_2, in rare gas clusters, $(I_2)Rg_N$ and molecular clusters, e.g. $(I_2)(OCS)_N$. Of particular interest are the number of solvent rare gas atoms or molecules required for cage recombination of the iodine radicals to occur. In these studies, the target chromophore molecule is electronically or vibrationally excited (or ionized) and the dynamics and kinetics of the dissociation and recombination processes are investigated. As an example of this work, Lineberger has found that the photodissociation of Br_2^- is suppressed (due to trapping and cage recombination) in CO_2 clusters, $(Br_2^-)(CO_2)_N$, with $N \gtrsim 10$.

The group of Müller-Dethlefs have made extensive use of zero kinetic energy electronic (ZEKE) spectroscopy to study hydrogen bonding and C–H$\cdots\pi$ and $\pi\cdots\pi$ interactions in small clusters containing aromatic molecules, solvated by other aromatic molecules, or by rare gas atoms or water molecules. The study of such interactions is crucially important in obtaining a better understanding of factors influencing, for example, the structures of nuclear base pairs in biological systems.

3.5.4 Solvation of Multiply Charged Transition Metal Ions

One of the many interesting features of transition metals is their ability to exhibit more than one charged (oxidation) state. On electrostatic grounds, the second ionization energy of a transition metal is considerably higher than the first. For example, for copper:

$$Cu \rightarrow Cu^+ \qquad\qquad I_1 = 7.72 \text{ eV}$$

$$Cu^+ \rightarrow Cu^{2+} \qquad\qquad I_2 = 20.29 \text{ eV}$$

Since the second IP of a transition metal is much higher than the first IP of ligand molecules, such as H_2O ($I_1 = 12.61$ eV) and NH_3 ($I_1 = 10.18$ eV), coordination of a single ligand molecule (L) to a doubly charged metal cation (M^{2+}) generally leads to $L \rightarrow M$ electron transfer (i.e. ionization of the ligand) and ligand decomposition:

$$\text{e.g. } Ni^{2+} + H_2O \rightarrow [Ni(OH)]^+ + H^+$$

For this reason, it has proven impossible to produce solvated doubly charged metal ions $M^{2+}(\text{Mol})_N$, by sequentially picking up neutral ligand molecules.

Stace and co-workers have circumvented this problem in an experiment where neutral metal atoms are picked up by pre-formed molecular clusters and the metal-ligand cluster is subsequently ionized by (100 eV) electron impact:

$$\text{e.g. } (H_2O)_N Ar_M + Cu \rightarrow Cu(H_2O)_N + M\,Ar$$
$$Cu(H_2O)_N + e^- \rightarrow [Cu(H_2O)_L]^{2+} + (N{-}L)\,H_2O + 3e^-$$

where the Ar atoms act as an energy sink, preventing water cluster fragmentation upon complexation of the metal atom. Partial fragmentation accompanies ionization and allows excess energy to be removed. In this non-equilibrium process, the double positive charge on the metal ion is stabilized by the presence of more than one ligand (i.e. solvent molecule), as there is now an energy barrier to $L \rightarrow M$ electron transfer.

In Stace's experiments, cluster ions are detected mass spectroscopically after approximately 10^{-4} seconds, by which time any unstable species will already have decayed. Magic numbers of solvent molecules can therefore be identified, as in the hexa-solvated Cu^{2+} ions $[Cu(H_2O)_6]^{2+}$ and $[Cu(NH_3)_6]^{2+}$, which are believed to represent the inner coordination shells of aqueous and ammonia solutions of Cu^{2+}. It should be noted, however, that the most intense MS peaks for both water and ammonia correspond to the species $[Cu(\text{Mol})_8]^{2+}$. Larger clusters readily lose water or ammonia molecules to generate the octa-coordinated species, which is itself stable towards such loss. The hexa-coordinated species, on the other hand, readily lose two solvent molecules, which is consistent with the Jahn–Teller tetragonal distortion of the octahedral d^9 Cu^{2+} ion in $[Cu(\text{Mol})_6]^{2+}$ species.

On the basis of these results and experiments on other metal ions, Stace and colleagues have proposed that the first coordination shell of M^{2+} ions, in the gas phase, corresponds to the coordination of eight, rather than six, solvent molecules, and that perhaps our notions about coordination and solvation in bulk solutions may also need to be reconsidered.

By making an analogy with the ring structures formed by small water and ammonia clusters (see previous discussion), Stace has suggested a likely structure for the $[Cu(\text{Mol})_8]^{2+}$ species to be a Cu^{2+} ion sandwiched between two hydrogen-bonded $(H_2O)_4$ or $(NH_3)_4$ rings, which adopt a staggered (square-antiprismatic) arrangement with respect to each other, as shown in Figure 3.9 for $[Cu(H_2O)_8]^{2+}$. For these octa-coordinate species, collision-induced fragmentation results in the loss of three or more molecules together, rather than individually. This is consistent with the postulated structure, which has intramolecular hydrogen bonds (ligand clustering) as well as solvent-ion interactions.

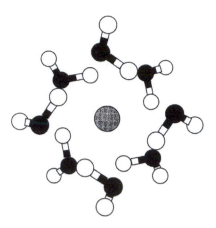

Figure 3.9 The square-antiprismatic structure proposed for $[Cu(H_2O)_8]^{2+}$.

Stace's experiments have shown that the minimum number of ligands required to stabilize Cu^{2+} is three for H_2O, but only two for NH_3, perhaps reflecting the greater dipole moment of the ammonia molecule. As mentioned above, the most intense peaks in the MS of $[Cu(Mol)_N]^{2+}$ occur at $N = 8$ for H_2O and NH_3; at $N = 6$ for ethanol; and at $N = 4$ for propanol and butanol. This decrease in preferred coordination number may reflect the increasing steric bulk of the ligands.

There is also interest in the electronic structures of ion-molecule clusters of this type. Stace and co-workers have investigated the *d–d*, ligand to metal charge transfer (LMCT) and metal to ligand charge transfer (MLCT) spectra of $[ML_N]^{2+}$ ions, where L is a 2-electron donor ligand molecule, such as H_2O, NH_3 or pyridine. These spectra, which are measured in the gas phase using photofragmentation spectroscopy, have enabled the comparison of the electronic structures of solvated ions in the gas phase with those in bulk solutions. In the future, this may enable the structures of gas phase ion–solvent clusters to be determined spectroscopically.

3.6 REACTIONS IN MOLECULAR CLUSTERS

In recent years there has been considerable interest in the study of reactions in molecular clusters, as models for reactions in atmospheric nanodroplets and in bulk solutions. A number of examples of molecular cluster reactions (especially those of water clusters) have already been discussed, but those presented here are intended to provide a brief overview of reactions of molecular clusters in general.

3.6.1 Cluster-Promoted Reaction

There are many examples where reactivity is initiated or promoted by clustering, and where the degree of clustering (i.e. the cluster size) influences the favoured reaction channel. As an example, while the nitric oxide cation does not react with a single water molecule, the cluster $NO^+(H_2O)_3$ undergoes the following bimolecular reaction with a further water molecule:

$$NO^+(H_2O)_3 + H_2O \rightarrow H_3O^+(H_2O)_2 + HNO_2$$

In this case, the reaction occurs at the stage of hydration where it first becomes exothermic to replace the NO^+ ion by H_3O^+ as the core of the cluster.

3.6.2 Ionization-Induced Reaction

Ionization, by electron bombardment, of CO_2 clusters generates excited cationic clusters, which undergo decomposition and loss of CO:

$$(CO_2)_N + e^- \rightarrow [(CO_2)_N^+]^* + 2e^-$$
$$[(CO_2)_N^+]^* \rightarrow [(CO_2)_{N-1}O^+]^* + CO$$
$$[(CO_2)_{N-1}O^+]^* \rightarrow [(CO_2)_{N-2}O_2^+] + CO$$

It should be noted that O_2^+ is created by the decomposition of two CO_2 molecules, as the reaction $CO_2^+ \rightarrow O_2^+ + C$ is too endothermic to be observed. The gas phase reaction corresponding to the above cluster reaction is:

$$O^+ + CO_2 \rightarrow O_2^+ + CO$$

An example of a negative cluster ion reaction, closely related to that above, is induced in N_2O clusters following electron capture:

$$(N_2O)_N + e^- \rightarrow [(N_2O)_N^-]^*$$
$$[(N_2O)_N^-]^* \rightarrow [(N_2O)_{N-1}O^-]^* + N_2$$
$$[(N_2O)_{N-1}O^-]^* \rightarrow [(N_2O)_{N-1}(NO)^-]^* + NO$$

Penning ionization results from the transfer of energy from an electronically excited atom or molecule (with excitation energy ΔE) to an atom or molecule with an ionization energy, $IP < \Delta E$. For example, collision between excited argon atoms and benzene molecules can lead to ionization of the benzene molecule:

$$Ar\ (3s^23p^6) + hv/e^- \rightarrow Ar^*\ (3s^23p^54s^1);\ \Delta E = 11\text{--}12\ eV)$$
$$Ar^* + C_6H_6\ (IP = 8.5\ eV) \rightarrow Ar + C_6H_6^+ + e^-$$

A cluster example of Penning ionization is given by:

$$Ar_N(C_2H_4) + e^- \rightarrow Ar^*Ar_{N-1}(C_2H_4) \rightarrow Ar_N(C_2H_4^+)^* + e^- \rightarrow Ar_N(C_2H_2^+) + H_2$$

where the reaction is driven by the energy difference between Ar^* and $C_2H_4^+$, with the excess energy leading to C–H bond breaking.

Ion molecule reactions often have high reaction rates, due to low activation barriers, and are responsible for many important processes, both in the earth's atmosphere and in interstellar space! In the ionosphere, the cation NO^+ is present in high abundance, due to photolysis of 'NOx' pollutants. It is believed that NO^+ is a nucleation site for the stepwise growth of small water clusters, up as far as the addition of three water molecules

$$NO^+ + 3H_2O \rightarrow (NO^+)(H_2O)_3$$

The next water molecule to be added, however, results in charge transfer (NO^+ acts as a charge sink) from NO^+ to H_2O, resulting in the fragmentation of a water molecule and the loss of nitrous acid, as shown in Section 3.6.1. An analogous cluster reaction, involving the collision-induced decomposition of $(NO^+)(H_2O)_4$, to yield $(H_3O)^+(H_2O)_2$ and HNO_2, has also been observed.

3.6.3 Cluster-Hindered Reaction

The opposite situation, where the presence of solvent molecules hinders or blocks a particular reaction channel is also known. Thus, the photodissociation of the CO_3^- anion:

$$CO_3^- + h\nu \rightarrow CO_2 + O^-$$

is blocked in small $(CO_3^-)(H_2O)_N$ clusters ($N = 1$–3), where the preferred reaction channel involves loss of water from the cluster.

3.7 FUTURE DEVELOPMENTS IN MOLECULAR CLUSTERS

There are almost limitless potential developments in the field of molecular clusters, mainly because of the diversity of possible molecular species. One important area is likely to be the use of size-controlled molecular clusters as nanolaboratories, in which to investigate fundamental reactions in a controlled manner, at the molecular level.

3.7.1 Biomolecular Clusters

Probably one of the most fascinating possibilities is the study of biomolecular clusters, or clusters of biophysically relevant molecules. Simons and co-workers have already used spectroscopic methods to investigate clusters of amino acids, solvated by water molecules. These workers have observed conformational changes which are induced by inter-molecular hydrogen bonding and have also noted the dependence of zwitterion formation ($^+H_3N-C(R_1)(R_2)-CO_2^-$) on the number of solvent molecules. Future studies in this area could centre on conformational studies of solvated polypeptides as models for *in vivo* proteins.

Jarrold and colleagues have extended their cluster ion-mobility studies (see Section 6.3.3) to include the investigation of the folding topologies of polypeptides and proteins, both in isolation, and in the presence of solvent water molecules, in order to establish the role played by solvent in determining protein folding and activity.

3.8 EXERCISES

3.1   Compare the intermolecular forces (and their relative strengths) in clusters of: nitrogen, benzene and water.

3.2   What is the evidence for small water clusters adopting structures which are not fragments of the bulk ice structure? Explain these findings.

3.9 FURTHER READING

Atkins, P.W., 2000, *Physical Chemistry*, 6th ed., (Oxford, Oxford University Press), Oxford), ch. 22.

Buck, U., 1994, Neutral molecular clusters. In *Clusters of Atoms and Molecules*, Vol.I, edited by Haberland, H., (Berlin, Springer-Verlag), pp. 374–395.

Castleman, A.W., 1994, Solvated cluster ions. In *Clusters of Atoms and Molecules*, Vol.II, edited by Haberland, H., (Berlin, Springer-Verlag), pp. 77–133.

Crofton, M.W., Price, J.M. and Lee, Y.T., 1994, IR spectroscopy of hydrogen bonded charged clusters. In *Clusters of Atoms and Molecules*, Vol.II, edited by Haberland, H., (Berlin, Springer-Verlag), pp. 44–76.

Leutwyler, S. and Bösiger, J., 1990, Rare-gas solvent clusters: spectra, structures and order-disorder transitions. *Chemical Reviews*, **90**, pp. 489–507.

Liu, K., Cruzan, J.D. and Saykally, R.J., 1996, Water clusters. *Science*, **271**, pp. 929–933.

Rigby, M., Smith, E.B., Wakeham, W.A., Maitland, G.C., 1986, *The Forces Between Molecules*, (Oxford: Oxford University Press).

Stone, A.J., 1996, *The Theory of Intermolecular Forces* (Oxford: Oxford University Press).

Sugano, S., 1991, *Microcluster Physics*, (Berlin, Springer-Verlag), pp. 118–132.

Wales, D.J., 1996, Structures, dynamics and thermodynamics of clusters: tales from topographic potential surfaces. *Science*, **271**, pp. 925–929.

CHAPTER FOUR

Metal Clusters I: Models

4.1 INTRODUCTION

In this chapter, I will present the theoretical models which have been developed to describe the bonding in clusters of metallic elements. It will be shown how these models can be used to explain various measured properties of metal clusters. The drawbacks and failings of these models are also emphasized. The variation of cluster structure (both geometric and electronic) with cluster size is also dealt with in this chapter. Further cluster experiments, and their interpretation, will be detailed in Chapter 5.

4.2 THE LIQUID DROP MODEL

The Liquid Drop Model (LDM) is the simplest model for metal clusters. It is a classical electrostatic model, in which the metal cluster is approximated by a uniform conducting sphere. The model has been used to derive scaling laws for the variation of a number of properties of metal clusters as a function of cluster size—as described in Section 1.2.4.

4.2.1 Ionization Energies and Electron Affinities

According to the LDM, the ionization energy (or ionization potential, IP) should decrease as the cluster gets larger (i.e. it requires less energy to remove an electron from a large cluster than from a smaller one). A detailed mathematical treatment leads to a linear dependence of the ionization potential on the reciprocal of the cluster radius, R:

$$\text{IP}(R) = W + \left(\frac{3}{8} \times \frac{1}{4\pi\varepsilon_0 R} \right) \tag{4.1}$$

where W is the work-function of the bulk metal (i.e. $\text{IP}(\infty)$). If the cluster radius, R is in Å and W is in eV, then Equation (4.1) becomes:

$$\text{IP}(R)/\text{eV} = (W/\text{eV}) + \frac{5.4}{(R/\text{Å})} \tag{4.2}$$

Such a linear dependence of IP on $1/R$ (or equivalently on $1/N^{1/3}$) has already been noted (in Section 1.2.4) for potassium clusters.

The LDM also predicts that the electron affinity (EA) of a cluster should increase as the cluster size increases (i.e. more energy is released on adding an electron to a large cluster than to a smaller one). Again there is a linear dependence of the electron affinity on the reciprocal of the cluster radius:

$$EA(R) = W - \left(\frac{5}{8} \times \frac{1}{4\pi\varepsilon_0 R}\right) \qquad (4.3)$$

If the cluster radius, R is in Å and W is in eV, then Equation (4.3) becomes:

$$EA(R)/\text{eV} = (W/\text{eV}) - \frac{9.0}{(R/\text{Å})} \qquad (4.4)$$

As the cluster tends towards the macroscopic size regime (i.e. as $N \to \infty$ and $1/R \to 0$) the ionization potential and electron affinity of the cluster should therefore both tend towards the bulk work-function, W. The predictions of the LDM for the ionization potential and electron affinity of metal clusters, as a function of the cluster radius R, are shown in Figure 4.1.

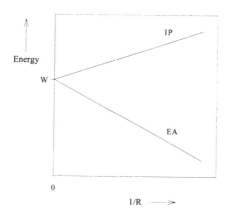

Figure 4.1 LDM prediction for the variation of ionization energy (IP) and electron affinity (EA) of metal clusters, as a function of radius (*R*)—see Equations (4.1) and (4.3).

By studying the photodetachment energies of a wide range of neutral and anionic metal clusters, Kappes has demonstrated that the LDM correctly predicts the (1/R) trends of decreasing IP and increasing EA with increasing cluster size— as can be seen in Figure 4.2. For small clusters (large values of 1/R), however, there are large deviations from the LDM predictions, which show up in Figure 4.2 as oscillations about the LDM straight line. As mentioned in Section 1.2.4, these oscillations are manifestations of the breakdown of the LDM at low nuclearity due to Quantum Size Effects (QSEs). The QSEs correspond to electronic shell closings, which will be discussed in Section 4.3, and are neglected by the classical LDM, in which there is no internal electronic structure (i.e. the cluster electrons are not quantized).

4.2.2 The HOMO-LUMO Gap

Photodetachment of electrons from anionic clusters (i.e. photoelectron spectroscopy) generates neutral clusters, which may be in excited electronic states:

$$M_N^- + h\nu \rightarrow [M_N]^* + e^-$$

By scanning through the frequency of the light, the energy level spectrum of the neutral cluster is mapped out. According to Koopman's approximation, the difference in energy between the highest occupied molecular orbital (HOMO) and the lowest unoccupied molecular orbital (LUMO) of the neutral cluster (ΔE) is approximately equal to the difference between the energies required to remove the highest energy and the second highest energy electrons of the anionic cluster.

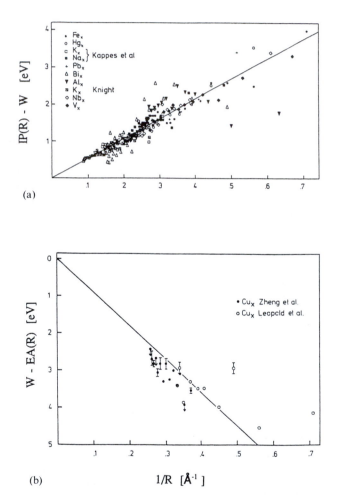

(a)

(b)

$1/R$ [Å^{-1}]

Figure 4.2 (a) Plot of experimental values of IP–W (W = work function) as a function of cluster radius (R) for a number of metallic clusters. (b) Plot of experimental values of W–EA as a function of cluster radius (R) for a number of metallic clusters. The straight lines represent the predictions of the LDM.

The energy of the HOMO of the neutral cluster is directly related to the ionization energy and the LUMO energy is related to the electron affinity of the cluster. Since, according to the LDM, both of these quantities vary as the cube root of the cluster nuclearity, the LDM predicts that the HOMO–LUMO gap ΔE is proportional to $N^{-1/3}$. As will be discussed further in Section 5.3, as clusters get larger their orbitals broaden into bands and the HOMO–LUMO separation becomes (in the language of solid state physics) the band gap, between the valence and conductance bands. On going from the cluster to the bulk metal ($N \to \infty$) ΔE tends to zero, as metals are characterized by vanishing band gaps. Experimentally, although the HOMO–LUMO gap is indeed found to decrease with increasing cluster size, the trend is not monotonous, due to electronic shell effects. In fact, photodetachment studies of small anionic copper clusters give a rough $N^{-1.8}$ dependence of ΔE on cluster size.

4.2.3 Surface Plasmons

While for small metal clusters, the electronic spectrum consists of a number of well defined lines corresponding to transitions between well separated, quantized energy levels, in medium-sized and large metal clusters, a single surface plasmon mode carries 100% of the oscillator strength so one observes a single peak in the electronic spectrum. The surface plasmon is due to extensive electronic correlation and corresponds to a collective excitation of conduction electrons relative to the ionic cores (as shown schematically in Figure 4.3)—i.e. the correlated motion of the cluster's itinerent electrons in the attractive field due to the positively charged ionic cores.

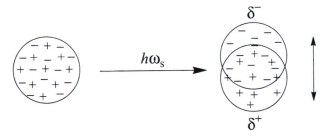

Figure 4.3 Schematic representation of the (dipolar) cluster surface plasmon (with resonant frequency ω_s), corresponding to the collective excitation of the valence electrons (–) relative to the ionic cores (+).

Mie developed a classical electrodynamical model, based on the LDM, to quantify the collective excitation of electrons in metallic particles. A similar approach (the Drude model) has also been applied to bulk metals. According to Mie theory, the frequency of the surface (or Mie) plasmon of a metallic particle (ω_s) is blue-shifted relative to the plasmon frequency of the bulk metal (ω_p) by an amount proportional to the inverse radius of the cluster. The surface plasmon line width is also inversely proportional to the cluster radius and thus becomes sharper as the cluster size increases. The experimental energy ($\hbar\omega_s$) and line width (Γ) of the surface plasmon resonance, for silver particles (with diameters 2 nm < D < 10 nm) embedded in an argon matrix, show a good agreement with the predictions of Mie theory (i.e. an inverse dependence on cluster diameter, and hence on cluster radius R):

$$\hbar\omega_s(D)/\text{eV} = 3.21 + \frac{0.58}{(D/\text{nm})} \tag{4.5a}$$

$$\Gamma(D)/\text{eV} = 0.04 + \frac{0.59}{(D/\text{nm})} \tag{4.5b}$$

The surface plasmon described here is strictly known as the dipolar surface plasmon, since it causes an electric dipole at the surface of the cluster. The dipolar plasmon is dominant at cluster radii $R \ll \lambda$ (the wavelength of the exciting radiation). There are other, higher multipole resonances but these are generally less intense than the dipolar plasmon.

It is interesting to note the similarity between the dipolar surface plasmon in clusters and the giant dipole resonance observed experimentally in atomic nuclei. Since the inter–nucleon interactions in nuclei are orders of magnitude stronger than the electron–ion interactions in clusters, giant dipole resonances in nuclei occur at excitation energies of the order of MeV (in the gamma-ray region of the spectrum), rather than eV (the UV-visible region). Similar plasmon modes (giant atomic resonances) were also predicted as a consequence of the early 'plum-pudding' model of the atom, due to Thomson.

While Mie theory works well for medium-sized and large clusters, where the electronic excitations are extensively correlated, it breaks down for small clusters. Thus, for sodium clusters with fewer than 50 atoms, there is generally more than one intense electronic absorption peak (apart from certain clusters with high symmetry, such as tetrahedral Na_8) and the $1/R$ dependence of peak widths is not maintained. These deviations arise due to quantum size effects, as discussed earlier. Small clusters, therefore, are best regarded as undergoing primarily one-electron excitations, with a limited amount of electron correlation.

4.2.4 Breakdown of the Liquid Drop Model

As has been mentioned above, while the Liquid Drop Model and related models (such as Mie theory) give a good qualitative and semi-quantitative description of the size-dependence of many properties of metal clusters, there are a number of deviations from the model.

First, for small clusters, there are significant deviations from the LDM-predicted $1/R$ ($N^{-1/3}$) dependence of ionization energies, electron affinities, and HOMO–LUMO gaps. The LDM cannot reproduce, for example, the even-odd alternation observed in the IPs of small sodium clusters. It also fails to explain the occurrence of the magic numbers—peaks of high intensity relative to neighbouring peaks—observed in the mass spectra of alkali metal, and other clusters. These deviations are due to quantum size effects, which cannot be treated by the classical LDM.

For many transition metals (such as iron and nickel), where ionizations involve the removal of quite tightly bound d electrons, only a small variation of the IP and EA are observed with changing cluster nuclearity. In these clusters, magnetic effects, due to electronic spin–spin interactions are also important.

Finally, as will be discussed in Chapter 5, the ionization energies of mercury clusters, when plotted against $1/R$, rather than following a simple LDM straight line exhibit a clear discontinuity, which has been attributed to a transition between non-metallic and metallic bonding.

These failures of the LDM indicate that, at least in the small cluster regime (and possibly for medium-sized clusters also), a new model is required, which explicitly takes into account the electronic structure of the cluster. A quantitative (and even a good qualitative) understanding of these clusters therefore requires a quantum mechanical treatment of the cluster, giving rise to discrete electronic states. Such a treatment is available in the jellium model.

4.3 THE JELLIUM MODEL

4.3.1 Magic Numbers

In a series of seminal experiments in the early 1980s, Knight and co-workers observed a periodic pattern of intense peaks in the mass spectra (MS) of alkali metal clusters. The nuclearities correponding to these peaks ($N = 2, 8, 20, 40, 58$, etc.) were called 'magic numbers' and were attributed to the enhanced stability of a cluster (corresponding to an intense peak) compared to its immediate neighbours—as discussed in Chapter 2 for rare gas clusters. Figure 4.4 shows the mass spectrum of sodium clusters, showing such magic number features. In contrast to the rare gas clusters discussed earlier, the magic numbers for these metal clusters were rationalized using the quantum mechanical (spherical) jellium model. It will be shown that the jellium model also explains the non-LDM behaviour of the IPs and EAs of small alkali metal clusters and many other deviations from simple scaling laws.

It should be noted that the jellium model was originally developed to explain the structures and stabilities of atomic nuclei. It has also been used to describe the electronic structure of the atom. The applicability of the jellium model over a wide range of length scales—from the inside of a nucleus via the electronic structure of an atom to that of a cluster of hundreds or even thousands of atoms—makes it a major unifying concept.

4.3.2 The Spherical Jellium Model

In the spherical jellium model, a metal cluster is modelled by a uniform, positively charged sphere filled with an electron gas. The Schrödinger equation is solved for an electron constrained to move within the cluster sphere under the influence of an attractive mean field potential due to the nuclei or ionic cores (whose actual positions are not important).

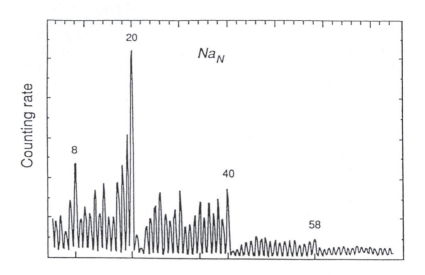

Figure 4.4 Mass spectrum of sodium clusters. Magic numbers are indicated above the relevant peaks.

It has been said that the neglect of the positions of the atomic nuclei in the jellium model is valid if the valence electrons are weakly bound and the ionic background responds easily to perturbations (i.e. when the delocalized valence electrons have *s*-wave character). Both of these conditions are met by the alkali metals and by the noble metals (copper, silver and gold).

A further condition, which remains a point of contention, is that the jellium model only applies when the clusters are molten. Since the melting points of clusters decrease as clusters get smaller and as bulk alkali metals have low melting temperatures, alkali clusters may indeed be molten under the experimental conditions pertaining to cluster creation and detection. In fact, theoretical calculations on Na_{20} do predict rapid structural isomerization at a temperature of 200 K. The low-temperature experiments of Whetten and co-workers, mentioned in Section 1.3.5.2, were said to show that magic numbers (jellium electronic effects) are not observed for cold, presumably solid-like clusters. However, it can be argued that the non-appearance of magic numbers is in fact due to the kinetic nature of the cluster production in these experiments and has nothing to do with size-dependent cluster stability, or its absence.

The jellium model is a quantum mechanical model with quantization of electron energy levels arising due to the boundary conditions imposed by the potential. This is in contrast to the classical Liquid Drop Model, where there is no electronic structure. The jellium model gives rise to electronic shell structure for clusters with up to several thousands of atoms. The jellium potential may be empirical or, alternatively *ab initio* effective potentials may be evaluated from Density Functional Theory (DFT) or Hartree–Fock molecular orbital calculations.

4.3.2.1 Ab Initio Jellium Model

In the self-consistent *ab initio* jellium treatment of Eckardt, electron-electron interactions are treated self-consistently using DFT (within the Local Density Approximation) for exchange and correlation. The model has a single parameter, the Wigner-Seitz radius (R_{WS}), which is the radius of a sphere whose volume is equal to the volume per atom (or per valence electron) in the solid. The jellium density (ρ_j) is given by:

$$\rho_j = \left(\frac{4\pi R_{WS}^3}{3} \right)^{-1} \tag{4.6}$$

and the electron density is determined variationally. In this way, a self-consistent effective Kohn–Sham potential can be defined, as shown for the case of Na_{40} in Figure 4.5.

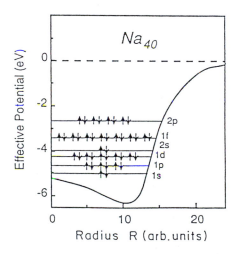

Figure 4.5 Self-consistent effective potential of jellium sphere corresponding to Na_{40}, with the electron occupation of the jellium energy levels included.

4.3.2.2 Empirical Jellium Models

Knight, Clemenger and co-workers developed empirical jellium models—based on assumed effective single particle potentials and solving the one-electron Schrödinger equation for a particle inside a sphere, constrained to move under the influence of an attractive central potential. It should be noted, however, that the results presented below apply equally well to jellium models based on *ab initio* potentials.

Due to the spherical nature of the potential and the fact that it acts from the centre of the cluster, the solutions of the Schrödinger equation (i.e. the jellium wavefunctions or orbitals) are separable into radial (R) and angular (Y) parts:

$$\psi_{n\ell m_\ell}(r,\theta,\phi) = R_{n\ell}(r) \cdot Y_{\ell m_\ell}(\theta,\phi) \tag{4.7}$$

where, by analogy with atomic quantum numbers, n, ℓ and m_ℓ are the principal, angular momentum and magnetic quantum numbers, respectively. Allowed values of these quantum numbers are as follows:

$n = 1, 2, 3 \dots$

$\ell = 0, 1, 2 \dots$ (no restriction on ℓ)

$m_\ell = -\ell \dots 0 \dots +\ell$ ($2\ell+1$) values

and jellium orbitals are labelled by analogy with atomic orbitals: $n = 1$, $\ell = m_\ell = 0$ corresponds to a 1s jellium orbital; $n = 1$, $\ell = 1$, $m_\ell = 0, \pm1$ corresponds to the set of three 1p jellium orbitals; etc. Magic numbers arise due to the complete filling of jellium levels or sub-shells (1s, 1p, 1d, 2s, 1f etc.).

It should be noted that the principal quantum number n in the jellium model, because of its origin in nuclear physics, is distinct to that used in the definition of atomic orbitals (n_{at}). The two are related by:

$$n = n_{at} - \ell \tag{4.8}$$

where ℓ is the angular momentum quantum number. Thus, the number of radial nodes in a jellium wavefunction is given by $n-1$ (as compared with $n_{at}-\ell-1$ for atomic orbitals).

In the spherical jellium model (as in the case of atomic orbitals) the energies of jellium orbitals increase with increasing principal quantum number (n) and increasing angular momentum quantum number (ℓ), while the ($2\ell+1$) orbitals with the same m_ℓ value are degenerate (have the same energy). The exact energy ordering of the jellium levels, however, depends on the radial form of the assumed effective jellium potential.

4.3.2.3 Types of Radial Jellium Potentials

(a) The 3-D harmonic potential is the simplest radial potential that has been employed within the spherical jellium model. This is the form of potential which would be expected for electrons moving in the field due to a smeared out positive charge, while ignoring electron–electron interactions. For the 3-D harmonic radial potential energy function:

$$U(R) = -k(R - R_0)^2 \tag{4.9}$$

all orbitals with the same value of ($2n+\ell$) are degenerate so the orbital energies can be expressed in terms of a single quantum number ($v = 2n+\ell-2$). This situation should be compared with the atomic orbitals of hydrogenic (single electron) ions, where the $1/r$ form of the Coulomb potential (and the lack of electron–electron screening) leads to a degeneracy of all atomic orbitals with the same value of the principal quantum number n_{at}.

The ordering of the jellium orbitals for the 3-D harmonic radial potential is: [1s (v=0)] < [1p (v=1)] < [1d, 2s (v=2)] < [1f, 2p (v=3)] < [1g, 2d, 3s (v=4)], etc. The total degeneracy of each set of orbitals is a triangular number (1, 3, 6, 10, 15, etc.), with a general set of levels (with quantum number v) having a degeneracy ½(v+1)(v+2). The predicted magic numbers for the 3-D harmonic potential are therefore: $N^* = 2, 8, 20, 40, 70$, etc. While, experimentally, magic numbers are found corresponding to 8, 20 and 40 electrons, this potential does account for the high relative intensity of clusters with 58 electrons.

(b) The 3-D square well potential, in which the potential is constant within the cluster sphere and is infinite at the classical cluster radius (R_0), has also been used for jellium calculations. This potential is predicted by a model which takes into account the exchange-correlation hole that exists around each electron in the cluster—i.e. the tendency for electrons to avoid each other due to coulomb repulsion and the Pauli exclusion principle. With the square well potential, there is zero electron density (zero probability of finding the electron) at a distance R from the cluster centre when $R \geq R_0$. This imposes the boundary condition on the radial jellium wavefunction that $R_{n\ell}(R)$ must be zero for $R \geq R_0$.

The radial wavefunction $R_{n\ell}(R)$ can be defined in terms of spherical Bessel functions $j_\ell(\kappa_{n\ell}R)$, for example:

$$j_0(\kappa_{n\ell}R) = \frac{\sin\kappa_{n\ell}R}{\kappa_{n\ell}R} \tag{4.10a}$$

$$j_1(\kappa_{n\ell}R) = \frac{\sin\kappa_{n\ell}R}{\kappa_{n\ell}^2 R^2} - \frac{\cos\kappa_{n\ell}R}{\kappa_{n\ell}R} \tag{4.10b}$$

where $\kappa_{n\ell}$ is a parameter which is related to the quantum numbers n and ℓ via the boundary condition $R_{n\ell}(R) = 0$ for $R \geq R_0$. For example, for the case of $\ell = 0$ (*ns* orbitals) κ has the form:

$$\kappa_{ns} = \frac{n\pi}{R_0} \tag{4.11}$$

The eigenvalues (jellium orbital energies) for the 3-D square well potential are given (in atomic units) by:

$$E_{n\ell} = \left(\frac{\hbar^2}{2m}\right)\kappa_{n\ell}^2 \tag{4.12}$$

As the energies of jellium orbitals depend on both n and ℓ, the orbital degeneracies are just $(2\ell+1)$—as for polyatomic atoms, although the ordering is not the same. The ordering of jellium orbitals, for the 3-D square well potential is $1s < 1p < 1d < 2s < 1f < 2p < 1g \ldots$, which gives rise to jellium shell closings (and therefore magic numbers) of $N^* = 2, 8, 18, 20, 34, 40$ and 58 (etc.) electrons. While there is no clear experimental evidence for a magic number corresponding to 34 electrons, this potential does correctly predict the stability of the 58 electron cluster.

(c) The Woods–Saxon potential is the most widely used (and most successful) radial potential in empirical jellium potentials. The potential, which was originally developed for the jellium description of atomic nuclei, has the following mathematical form:

$$U(R) = \frac{-U_0}{\exp[(R - R_0)/\sigma] + 1} \tag{4.13}$$

where

$$U_0 = E_F + W \tag{4.14}$$

(E_F and W are the Fermi energy and the work-function of the bulk metal, respectively)

$$R_0 = R_{WS} N^{1/3} \tag{4.15}$$

(R_0 is the effective radius of the cluster sphere and R_{WS} is the Wigner–Seitz radius) and σ is a constant scaling distance of 1.5 bohr ($\approx 7.94 \times 10^{-11}$ m).

As shown in Figure 4.6, the Woods–Saxon potential is a finite, almost-square well with rounded sides and is thus intermediate in shape between the 3-D harmonic potential and the 3-D square well potential discussed above. This reflects the fact that the interactions in jellium-like clusters are intermediate between the extremes represented by the harmonic and square well potentials. The shape of the Woods–Saxon potential is also similar to the effective Kohn–Sham potential, derived from *ab initio* calculations, which was shown in Figure 4.5.

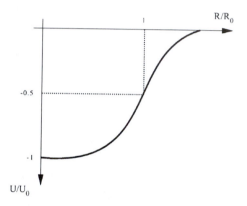

Figure 4.6 Sketch of the Woods–Saxon empirical jellium potential.

Figure 4.7 compares the jellium orbital orderings of the Woods–Saxon potential with those of the 3-D harmonic and square well potentials. It is not surprising to see that the Woods–Saxon potential has an orbital energy ordering which is intermediate between those of the simpler potentials. As for the square well potential, the orbital energies depend on both n and ℓ, so the orbital degeneracies are correct. For small clusters, the jellium level ordering: $1s < 1p < 1d < 2s < 1f < 2p < 1g$ and shell closings for $N^* = 2, 8, 18, 20, 34, 40, 58$ etc. are the same as for the 3-D square well potential. The first difference in orbital ordering between the Woods–Saxon and square well potentials occurs above the $2d$ orbital (i.e. for more than 68 electrons).

4.3.2.4 Correlation of Cluster Properties with Jellium Model Predictions

(a) Mass spectral abundances. As mentioned above, the Woods–Saxon potential, within the jellium model, explains the occurrence of the magic number peaks in the mass spectra of alkali metal and noble metal clusters (both with one valence electron per atom). If the clusters are generated and ionized under conditions,

which ensure that the MS intensities reflects the abundance of neutral clusters, then magic numbers are observed at nuclearities $N^* = 2, 8, 18, 20, 34, 40, 58$, etc. However, if the experimental conditions are such that MS intensities reflect the stabilities of the cations, then magic numbers occur at nuclearities which are increased by one relative to the neutral clusters: $N^* = 3, 9, 19, 21, 35, 41, 59$, etc. For example, M_{21}^+ has a magic number (20) of electrons in the jellium model. Similarly, under conditions where MS intensities reflect the stabilities of anionic clusters, then magic numbers occur at nuclearities which are one less than the neutral clusters: $N^* = 7, 17, 19, 33, 39, 57$, etc. For example, M_{33}^- has a magic number (34) of electrons in the jellium model.

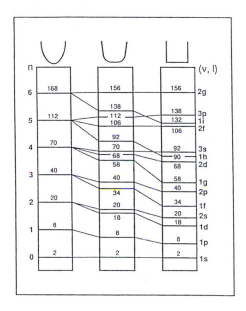

Figure 4.7 Comparison of energy level orderings and degeneracies for spherical 3-D, harmonic (left), intermediate (Woods–Saxon, centre), and square well (right) potentials. The numbers above the energy levels indicate cummulative totals of electrons.

The mass spectra of certain divalent metals such as zinc and cadmium can also be explained by the jellium model. For a divalent metal, assuming that the MS intensities reflect neutral cluster stability, magic numbers would be expected to occur at nuclearities which are half those of the monovalent metals ($N^* = 4, 9, 10, 17, 20, 29$, etc.), as the number of cluster (jellium) electrons $N_e = 2N$. The mass spectra of divalent zinc and cadmium clusters are reproduced in Figure 4.8.

For trivalent metals, such as aluminium, the jellium model does not show such good agreement with experimental mass spectra, perhaps due to the electron-localizing effect of the high (+3) ionic core charges and incomplete *s-p* hybridization for small Al clusters, though there is evidence that the jellium model can account for magic numbers of Al$_N$ clusters in the range $N \approx 46$–306 atoms ($N_e \approx 138$–1,218 electrons).

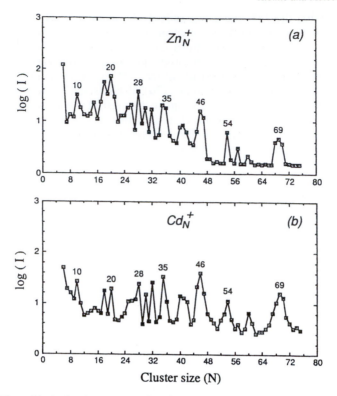

Figure 4.8 Logarithmic abundance spectra of (a) zinc and (b) cadmium clusters. Numbers of atoms are indicated corresponding to closed jellium shells ($N_e = 2N$).

(b) Ionization energies. Experiments on alkali metal clusters reveal that the ionization energies (IPs) of stable closed jellium shell clusters are high and that there is an abrupt drop in IP immediately after the closing of a jellium shell. The enhanced stability associated with a magic number ($N_e = N$ for neutral clusters) of electrons, corresponding to complete filling of a set of jellium orbitals, leads to a high IP for M_N, while the IP of the cluster M_{N+1}, with one extra electron, will be low. Jellium shell effects are superimposed on the $N^{-1/3}$ (or R^{-1}) trend of the LDM, to give the oscillatory pattern of IPs mentioned earlier. The even–odd alternations (shown in Figure 4.9 for potassium clusters) are due to spin-pairing effects and will be discussed below. Finally, it is also known that the photoionization cross sections of closed jellium shell clusters are lower than for open shell clusters. This has significance for mass spectral measurement made using near-threshold ionization, when the lower photoionization cross sections lead to dips, rather than peaks, in intensity for closed shell clusters.

(c) Electron affinities. Experimental measurements of electron affinities (EAs) of copper clusters are in good agreement with the predictions of the jellium model. Thus, neutral clusters with closed jellium shells have low electron affinities, while those with open partially filled jellium orbitals have high EAs. The electron

affinities have been obtained by measuring electron photo-detachment energies of the corresponding anionic clusters, since the energy required to remove an electron (adiabatically) from Cu_N^- is, by definition, the electron affinity of the neutral cluster Cu_N. Using the argument developed above for IPs, if the N-electron neutral cluster Cu_N corresponds to a filled jellium shell, the $(N+1)$-electron anionic cluster Cu_N^- must have an open shell and (as discussed above for IPs) a low detachment energy—corresponding to a low EA for the neutral cluster.

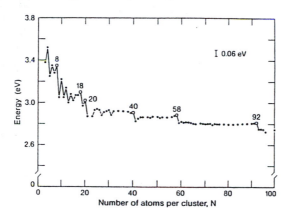

Figure 4.9 Plot of ionization energies of potassium clusters as a function of N. Numbers indicate jellium shell closings.

(d) Electronic polarizabilities. de Heer and co-workers have measured the polarizabilities of aluminium clusters with $N = 15$–61 atoms. Polarizabilities are measured by passing a collimated beam of neutral Al clusters through an inhomogeneous electric field. The clusters are deflected by the electric field, with the deflection being proportional to the cluster polarizability (α_{clus}) and inversely proportional to its mass, and hence the number of atoms in the cluster.

From Mie theory, treating the cluster as a spherical liquid drop, the polarizability per atom (α_{LD}) of the cluster is given by:

$$\alpha_{LD} = \frac{\alpha_{clus}}{4\pi\varepsilon_0 N} = \frac{R^3}{N} \tag{4.16}$$

while jellium models tend to give larger polarizabilities, corresponding to an increased effective cluster radius:

$$\alpha_j = \frac{(R+\delta R)^3}{N} \tag{4.17}$$

where $\delta R = 1.5$ bohr. Since the cluster volume $V \propto R^3 \propto N$ (see Section 1.2.1), the polarizability per atom ($\alpha_{clus}/N \propto R^3/N$) should be size-independent.

Experiments on Na and K clusters show a slight decrease in α_{clus}/N with increasing N, with superimposed dips in polarizability for closed shell clusters, (e.g. $N = 8$ and 20), where the electrons are more tightly bound. The polarizabilities for Na and K clusters are almost identical as the ratio R_{WS}^3/α_{atom} is very similar for the two metals. The experimental electronic polarizabilities (per atom) of aluminium clusters (in the range $N = 15$–61) increase slightly with increasing N, but there are again large size-dependent oscillations, which may be due to jellium shell closings. The low polarizabilities measured for small Al clusters may be due to lower s-p mixing in this size regime, which would be consistent with the evidence from MS measurements.

(e) Magnetism. The magnetism of metal clusters has been measured by performing Stern–Gerlach type experiments (deflection of a cluster beam in an inhomogeneous magnetic field) and by performing electron spin (paramagnetic) resonance of lithium and aluminium clusters in inert gas matrices. These experiments have shown that simple metal clusters (those of the alkali metals, the noble metals and aluminium) have electronic structures corresponding to the pairwise occupation of jellium orbitals, so that odd-electron clusters are paramagnetic (with only one unpaired electron) and even-electron clusters are generally diamagnetic (with no unpaired electrons).

It should be noted that, according to Hund's rule of maximum multiplicity as used for predicting the ground states of atoms, the simple spherical jellium model predicts that clusters with incompletely filled degenerate jellium orbitals should have the maximum number of unpaired electrons, so as to maximize the exchange energy. Even clusters with an even number of electrons, but incomplete filling (e.g. $1s^2 1p^2$) should have high-spin ground states. The fact that most even-electron simple metal clusters are actually diamagnetic arises because the clusters can undergo low-energy Jahn–Teller distortions that remove some of the orbital degeneracy and lead to electron pairing. This will be discussed further in Section 4.2.3.

As we shall see in Section 5.3, transition metal clusters differ from simple metals in that they can have large numbers of unpaired electrons and (as in the cases of Fe, Co and Ni) they can exhibit temperature-dependent magnetic ordering and superparamagnetism.

(f) Reactivity studies of simple metal clusters have revealed that closed jellium-shell clusters, such as the 20-electron cluster, Al_7^+, have lower reactivity towards small molecules than do open shell clusters. The stability of 20-electron clusters of aluminium is emphasized by the fact that the reaction between Al_N and H_2 is only exothermic for Al_6, as Al_6H_2 has 20 valence electrons.

No such patterns are observed for transition metal clusters, where reactivities (see Section 5.2) tend to be more related to geometric, rather than electronic factors.

4.3.2.5 Breakdown of the Spherical Jellium Model

We have seen that for understanding and predicting the physical properties of small metal clusters, the jellium model, which takes into account the internal electronic structure of the cluster, is a significant improvement over the Liquid Drop Model. However, in the previous section, we have encountered a number of cases where the simple spherical jellium model does not satisfactorily explain all of the observables, notable examples being the fine structure (even–odd alternations) observed in the mass spectra, IPs, EAs and polarizabilities of simple metal clusters and the diamagnetism of even-electron clusters with formally open jellium shells.

4.3.3 Perturbed Jellium Models

The deficiencies of the spherical jellium model can be overcome by allowing non-spherical cluster geometries, which leads to a reduction of the jellium orbital degeneracies.

4.3.3.1 The Ellipsoidal Shell (Clemenger-Nilsson) Model

In 1985 Clemenger developed the ellipsoidal shell model, which was based on Nilsson's model (dating from 1955) of ellipsoidally distorted atomic nuclei. In the ellipsoidal shell model, ellipsoidal cluster distortions are allowed for partially filled jellium orbital shells. The potential adopted is a perturbed 3-D harmonic oscillator, with distinct force constants (k_x, k_y and k_z) along the three cartesian axes of the ellipsoid, though in the basic ellipsoidal shell model axial symmetry is maintained, such that at least two of the force constants are always equal.

The lowering of the spherical symmetry results in the loss of the $(2\ell+1)$ degeneracy of each jellium $n\ell$ shell. Keeping axial symmetry means that the $\pm m_\ell$ degeneracy is maintained, so that the $(2\ell+1)$ orbitals split into ℓ ($\pm m_\ell$) pairs and the $m_\ell = 0$ orbital is left on its own. As shown in Figure 4.10a, two types of ellipsoidal distortions are possible: *oblate* and *prolate*.

An oblate ellipsoid has two equal long axes (x,y) and one shorter axis (z). The moments of inertia about the three axes are thus in the order $I_x = I_y < I_z$. In the ellipsoidal shell model, the oblate geometry has 3-D harmonic force constants $k_x = k_y > k_z$, so that electronic motion along the z-axis is more constrained than in the xy plane. In terms of a simple particle in a box model, the shorter the box length the higher the energy level. This results in the energies of jellium orbitals being highest for those orbitals with greatest amplitude parallel to the z-axis. Thus, the orbital energies decrease as $|m_\ell|$ increases, as shown (for a set of np orbitals) in Figure 4.10b.

The situation for the prolate ellipsoid, which has two equal short axes (x,y) and one longer axis (z), is the reverse of the oblate ellipsoid. The 3-D harmonic force constants are in the order $k_x = k_y < k_z$ and the energies are lowest for those orbitals with greatest amplitude parallel to the z-axis (i.e. orbital energies increase as $|m_\ell|$ increases) as shown in Figure 4.10b.

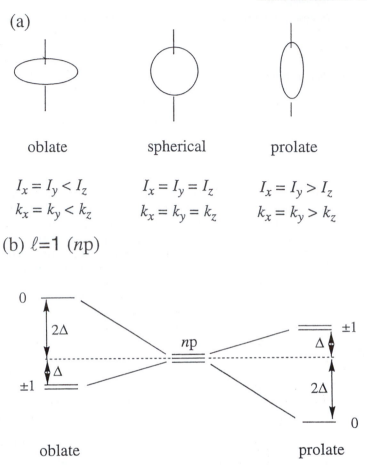

Figure 4.10 The ellipsoidal shell model. (a) Ellipsoids and their relative moments of inertia and harmonic force constants. (b) Splitting of spherical jellium *np* orbitals upon ellipsoidal distortion.

The complementary nature of the jellium orbital splitting, which accompanies prolate and oblate distortions from spherical symmetry, has been used to explain the shapes of clusters with incomplete filling of jellium shells. Thus, clusters with up to half-filled shells (e.g. p^1–p^3 or d^1–d^5 configurations) are predicted to have prolate geometries. Beyond the half-filled shell there should be an abrupt change to oblate geometry, which persists until the shell is filled. These predictions are in good agreement with geometries calculated using high quality *ab initio* molecular orbital methods, by Bonačič-Koutecký and others.

4.3.3.2 Other Non-spherical Jellium Models

Though the ellipsoidal shell model, as implemented by Clemenger, is an empirical model, a number of self-consistent calculations have been carried out, based on this model.

An alternative way of accounting for the occurrence of non-spherical clusters is that the partial filling of degenerate jellium orbitals gives rise to a degenerate ground electronic state which is unstable with respect to a symmetry-lowering distortion—analogous to the Jahn–Teller distortion observed in molecules and crystals. This descent of symmetry approach has been extended independently, by Upton and Mingos, to include a crystal field-like splitting of the spherical jellium shells under the actual symmetry of the cluster, as defined by the nuclear positions. With such models, however, the inherent simplicity of the jellium model is lost.

4.3.3.3 Ab Initio Calculations

More detailed models of the electronic structures of metal clusters include semi-empirical and *ab initio* molecular orbital (MO) calculations and the application of Density Functional Theory (DFT). Bonačič-Koutecký and co-workers, in particular, have used *ab initio* MO theory (including configuration interaction) to rationalize photoelectron spectra and depletion spectroscopy (see Chapter 1) of alkali metal and silver clusters, in order to go beyond the simple jellium model and obtain the actual arrangement of the atomic nuclei in small metal clusters.

4.4 BEYOND THE JELLIUM MODEL: ELECTRONIC SHELLS, SUPERSHELLS AND BANDS

So far we have seen how the jellium model explains the mass spectra of clusters with a few tens of atoms. But what about larger clusters? In 1991, Bjørnholm measured the mass spectra of sodium clusters with up to 600 atoms and observed spherical jellium shell closings throughout this size range. Around this time, Martin measured the mass spectra of sodium clusters up as far as 25,000 atoms and observed two series of periodic oscillations in intensity which were approximately evenly spaced when plotted against $N^{1/3}$, as can be seen from Figure 4.11. I will start by discussing clusters with fewer than 2,000 atoms – clusters with more than 2,000 atoms will be discussed in Section 4.5.

4.4.1 Electronic Shells

For $N < 2,000$, the oscillations in MS intensity have a small period. The magic numbers ($N^* \approx 340$, 440, 560, 700, 840, 1,040, 1,220, 1,430) are indicated by dips in mass spectral (MS) intensity. In these experiments, near-threshold photo-ionization is used to produce the cluster cations and the more stable clusters tend to have higher IPs and therefore smaller ionization cross sections.

The magic numbers in this size regime have been attributed to the filling of electronic shells, which occur due to the bunching together of jellium electronic energy levels (sub-shells) at high energy. For high energies (i.e. high values of n and/or ℓ) it has been shown that jellium levels with the same value of the quasi-quantum number ($n_{shell} = 3n + \ell$) are approximately degenerate. For example, the following sets of jellium orbitals are approximately degenerate:

$1k(n = 1, \ell = 8); 2h(2,5); 3d(3,2)$ $3n+\ell = 11$

$1q(1,13); 2m(2,10); 3j(3,7); 4g(4,4); 5p(5,1)$ $3n+\ell = 16$

An analogy can be made between the quasi-degeneracy of cluster jellium orbitals with the same values of $(n_{shell} = 3n+\ell)$ and the hydrogen atom, for which orbitals with the same value of n_{at} (i.e. the same $n+\ell$) and the 3-D harmonic osciallator, for which orbitals with the same value of $(2n+\ell)$ are degenerate. A semi-classical interpretation of the $(3n+\ell)$ quasi-quantum number, based on this analogy, involves the mixing of high-energy jellium wave functions, with the same $(3n+\ell)$ value to generate classical wave packets, which correspond to classical closed triangular paths or orbits

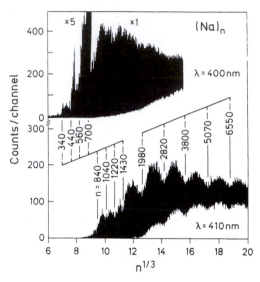

Figure 4.11 Periodic oscillations in mass spectral intensity measured for large sodium clusters. Note that at lower photon energies, the magic number dips are more pronounced.

4.4.2 Electronic Supershells

In addition to oscillations due to electronic shell structure, Martin's experiments (Figure 4.11) revealed longer period intensity variations (i.e. periodic appearance and disappearance of shell structure, as at $N \approx 1,500$) in the region $N < 2,000$. This pattern can be understood in terms of the merging of electronic shells into dense band-like blocks or supershells. A semi-classical rationalization of this periodic oscillation has been put forward, which invokes interference between classical closed triangular and square electronic orbits, resulting in a beat pattern.

4.4.3 Shell Energy

The total electronic binding energy (a negative quantity) of a cluster, at the one-electron level (i.e. ignoring electron-electron interactions) is merely the sum of the individual electron binding energies:

$$E(N) = \sum_{i=1}^{nN} \varepsilon_i \qquad (4.18)$$

for a neutral cluster, with n electrons per atom. If the electronic energies are defined relative to those of a free atom, then the cluster cohesive energy can be obtained as the negative of the binding energy.

From the LDM, the cluster electronic binding energy is given by:

$$E_{LD}(N) = -aN + bN^{2/3} \qquad (4.19)$$

where a and b are constants, the first term gives the volume (bulk) contribution to the electronic binding energy and the second gives the correction due to the presence of the cluster surface. In terms of the binding energy per atom, this gives:

$$E_{LD}(N)/N = -a + bN^{-1/3} \qquad (4.20)$$

Considering the effect of electron quantization, within the jellium model, one can define the electronic shell energy as the difference between the electronic binding energy (calculated by from the jellium model, $E_j(N)$) and that from the LDM:

$$E_{shell}(N) = E_j(N) - E_{LD}(N) \qquad (4.21)$$

As can be seen in Figure 4.12, $E_{shell}(N)$ exhibits oscillations which are regularly spaced on a $N^{1/3}$ scale. At low nuclearities, these oscillations are due to the filling of individual jellium orbitals. Closed shells correspond to large negative values of $E_{shell}(N)$, since $E_j(N) < E_{LD}(N)$. Open shells, on the other hand give rise to positive values of $E_{shell}(N)$. At higher nuclearities, the oscillations are due to the filling of the quasi-degenerate electronic shells. The overall, large period oscillation is due to the supershell beat pattern.

Assigning an index s to each electronic shell, it can be shown that the total number of electrons required to fill electronic shells s up to a certain value (S) is given by:

$$N(S) = \sum_{s=0}^{S} \sum_{\ell=0}^{L(s)} 2(2\ell+1) = aS^3 + bS^2 + cS + d \approx aS^3 \qquad (4.22)$$

where $L(s)$ is the highest angular momentum subshell in shell s. It is apparent that the magic numbers, corresponding to filled electronic shells, N^* are approximately proportional to S^3—i.e. $(N^*)^{1/3} \propto S$. The electronic shell magic numbers (oscillations in E_{shell} and in the MS intensities) are linearly dependent on the outer shell value, S, which explains why they are evenly spaced when plotted against $N^{1/3}$. Bjørnholm has shown that, within the jellium model, using the Woods–Saxon potential, the constant a in Equation (4.22) has a value of approximately 0.21.

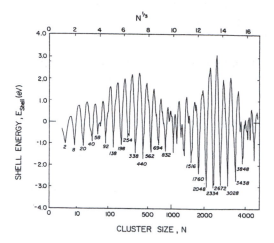

Figure 4.12 Calculated shell energies (E_{shell}) for spherical sodium clusters.

In fact, Equation (4.22) holds for any spherically-symmetric potential and the corresponding a values for the Coulomb and 3-D harmonic oscillator potentials are ⅔ and ⅓, respectively.

4.5 GEOMETRIC SHELL STRUCTURE

4.5.1 Geometric Shell Structure in Alkali Metal Clusters

For sodium clusters with over 2,000 atoms (see Figure 4.11), Martin and co-workers observed long period oscillations in intensity with magic numbers ($N^* \approx$ 1,980, 2,820, 5,070, 6,550,..., 18,000, 21,300) again appearing as dips in MS intensity (see Figure 4.11). In this size regime, the magic numbers have been ascribed to the filling of concentric polyhedral, or *geometric shells* of atoms, rather than electrons. As mentioned in Chapter 2, filled geometric shells, which impart stability to the cluster by maximizing the average coordination number and thereby reducing the cluster surface energy, were first detected in the mass spectra of rare gas clusters. For sodium clusters (and also for other alkali metal clusters in this size regime), the observed magic numbers are close to those expected for geometric shell clusters, based on twelve-vertex polyhedra, such as the icosahedron, the Ino decahedron or fcc-like cuboctahedron. Magic numbers appear as dips in MS intensity in near-threshold ionization experiments because filled geometric shells have higher ionization energies than incomplete shells, where unsatisfied coordination leads to localization of electrons in surface states.

Though they are characterized by the same magic numbers for complete geometric shells, by analysing the fine structure in the MS intensity plots, it is possible to distinguish between icosahedral, decahedral and cuboctahedral growth, since these structures have different secondary magic numbers associated with geometric sub-shells. Geometric sub-shells result from the covering of individual faces or groups of faces of a polyhedron without giving rise to a complete polyhedral shell. In this way, Martin determined that sodium (and other alkali metal) clusters in the size range 2,000–25,000 follow an icosahedral growth pattern.

4.5.2 Geometric Shell Magic Numbers

As discussed in Chapter 2 for rare gas clusters, geometric shell clusters based on twelve-vertex polyhedra (e.g. icosahedra, decahedra and cuboctahedra) and consisting of K concentric shells are characterized by the following magic numbers (N^*):

$$N^*(K) = \frac{1}{3}\left(10K^3 + 15K^2 + 11K + 3\right) \tag{4.23}$$

i.e. N^* = 13, 55, 147, 309, 561, 923, 1,415, 2,057, 2,869, 3,871, 5,083, 6,525, 8,217,..., 21,127, etc. (Note that the shell index K is used to refer to geometric shells, in contrast to S, which refers to electronic shells.) Five-shell polyhedral clusters ($K = 5$, $N^* = 561$) with these geometries are shown in Figure 4.13.

Analogous magic numbers for bcc-like clusters, based on the fourteen-vertex rhombic dodecahedron, are given by:

$$N^*(K) = 4K^3 + 6K^2 + 4K + 1 \tag{4.24}$$

i.e. N^* = 15, 65, 175, 369, 671, 1,105, 1,695, 2,465, 3,439, 4,641, etc. A five-shell (671-atom) rhombic dodecahedral cluster is shown in Figure 4.13.

4.5.3 Geometric Shell Structure in Clusters of Other Metals

4.5.3.1 Alkaline Earth Clusters

Mass spectral studies of clusters of the alkaline earth metals (the group 2 elements Be, Mg, Ca, Sr and Ba) have shown that, as for the large alkali metal clusters, they adopt icosahedral shell geometries. However, in contrast to the alkali metals, geometric shell structure is seen even for small clusters (of a few tens of atoms). Figure 4.14a shows the mass spectrum of barium clusters, showing (as for the rare gases) significant features at $N = 13$ and 19. The upper limits to the icosahedral structures have not yet been determined: thus magnesium is icosahedral up to 3,000 atoms, calcium and strontium up to 5,000—as shown in Figure 4.14b for calcium clusters.

The icosahedral growth pattern observed for the alkaline earth metals, for all cluster sizes studied, is analogous to that found for the rare gases. This analogy occurs because the group 2 atoms have formally closed sub-shell $(ns)^2(np)^0$ electron configurations—whereas rare gas atoms have closed shell $(ns)^2(np)^6$ configurations—and so electronic effects are small.

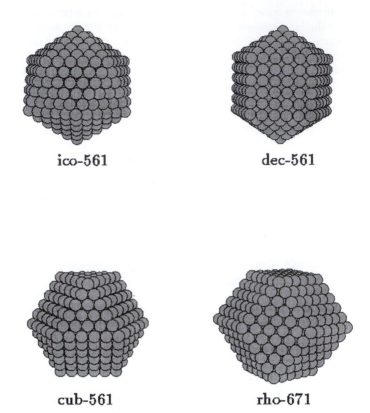

Figure 4.13 The outer shells of five-shell geometric shell metal clusters: ico = icosahedron; dec = (Ino) decahedron; cub = cuboctahedron; rho = rhombic dodecahedron. The total number of atoms in each cluster is also indicated.

For larger alkaline earth clusters, the bands of orbitals derived from the *s* and *p* atomic orbitals must overlap, thereby making the clusters metallic. However, at these nuclearities, electronic shell structure is not manifest in the MS because the level spacings are already small enough to wash out any electronic preferences and geometric shell structure is again observed.

The analogy between rare gas atoms and metal atoms with $(ns)^2(np)^0$ valence electron configurations also explains the size-dependent variation of properties of mercury clusters, as will be discussed in Chapter 5.

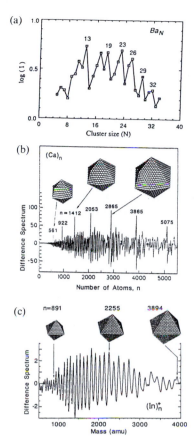

Figure 4.14 Mass spectra of clusters of some main group metals: (a) Ba; (b) Ca; (c) In.

4.5.3.2 Group 13 Clusters

Aluminium clusters in the range ($46 < N < 306$) show jellium shell closings but the situation for smaller clusters is more ambiguous, perhaps due to incomplete *s-p* hybridization meaning that the three valence electrons cannot be regarded as equally delocalized. It has been suggested that the simple jellium model should be less successful for high formal charges on the ionic core. The $+3$ charge of the Al core is expected to perturb the jellium electrons by increasing the localized ion–electron attraction. For small clusters, this will also increase the covalency and directionality of the metal-metal bonding, compared with that found in alkali metal clusters. Finally, because of the greater number of accessible valence orbitals of Al (compared with the alkali metals) the density of electronic states rises more quickly with N than for alkali metal clusters and the geometric shell regime is reached at considerably smaller nuclearities.

The MS of large aluminium clusters, with 1,000–10,000 atoms (i.e. with 3,000–30,000 electrons) exhibits oscillations, which have been attributed (as discussed above for sodium clusters) to geometric shell closings. In the case of aluminium (and also its heavier group 3 congener indium), the geometric shells are octahedral, with local face centred cubic (fcc) packing (see Figure 14.14c for the mass spectrum of indium clusters), as in bulk aluminium.

4.5.3.3 Transition Metal Clusters

Clusters of the transition metals often have mass spectra characteristic of geometric shell structure, even for clusters as small as tens of atoms. This is because the partially filled set of five tightly bound valence d orbitals on each atom give rise to narrow bands, with a high density of electronic states, and therefore small energy level separations. Thus, electron shell effects are dominated by packing (geometric shell effects) even for low cluster nuclearities

For nickel and cobalt, Pellarin *et al.* have observed dips in MS intensity (corresponding to magic numbers in these near-threshold ionization studies) at N^* = 55, 147, 309, 561 and 925 atoms. Analysis of the fine structure of the MS has led to the conclusion that, in this size regime, Ni and Co clusters grow in an icosahedral fashion.

The situation for iron is more ambiguous. Pronounced troughs in the MS are observed at $N^* = 66, 88, 116, 147, 174$ and 206 and these magic numbers do not conform simply to any one of the usual geometric shell magic numbers. It has been suggested that, in the case of Fe clusters, there may be competition between electronic (jellium) and geometric shell effects: for example $N = 147$, corresponds to a magic number for a three-shell icosahedral, decahedral or cuboctahedral cluster. However, another possibility is that there is actually a competition between icosahedral (or decahedral or cuboctahedral) and rhombohedral (bcc) growth, as the MS dips at $N^* = 66$ and 174 are very close to the magic numbers for two-shell ($N^*=65$) and three-shell ($N^*=175$) rhombic dodecahedral clusters, respectively. Rhombohedral growth is consistent with the bcc structure of the low temperature allotrope of iron in the solid state (in contrast to Ni and Co, which adopt fcc packing in the solid state).

4.5.4 Competition Between Electronic and Geometric Shell Structure

The size at which cluster stability becomes governed by geometric, rather than electronic effects depends on a number of factors, such as the electronic density of states (DOS), atomic electron configuration, cluster-melting temperature and the temperature of the cluster. These factors are also critical for the size-dependent metal-insulator transition in clusters of metallic elements, which will be discussed in detail in Section 5.3.

It has been said that, under experimental conditions, the atoms (nuclei) in small clusters are mobile, so that the cluster can be regarded as a spherical liquid-like droplet. Quantization of the electrons' motions leads to jellium electronic structure. As the clusters get larger, the jellium levels bunch into electronic shells and supershells and eventually into bands where the level spacing is quasi-continuous (i.e. level spacing $\Delta \ll kT$). This is clearly just the opposite process to that of taking a metallic particle and decreasing its size until the electronic energy levels become discrete.

It is reasonable to suppose that, when the level spacing Δ is small compared to kT, the pattern of experimental MS intensities will no longer be governed by electronic shell filling, since there will be no preference for a certain electron count over any other. If the temperature is below the melting temperature (T_m) of the cluster (which rises with increasing cluster size) the clusters are solid-like and the clusters grow as nanocrystallites. Structures and stabilities are governed by crystal growth effects (such as the minimization of surface energies) and geometric, rather than electronic shell structure is observed.

The temperature of the cluster in relation to its melting temperature (T_m) is of critical importance in deciding whether electronic or geometric effects prevail. For $T > T_m$, the cluster is liquid-like and resembles a spherical liquid drop. Geometric shell structure is observed to disappear upon cluster melting since spherical clusters are not regular polyhedra and so there is no inherent stability associated with any given nuclearity. Martin and co-workers have used the disappearance of MS features related to geometric shell structure as a measure of the cluster melting temperature. In certain cases, such as sodium and aluminium cluster melting is actually accompanied by a transition from geometric to electronic shell structure.

Lermé and co-workers have shown that gallium clusters exhibit electronic shell structure even up to 5,000 atoms, indicating that they are presumably liquid-like over a wider range of temperatures than aluminium clusters. This is consistent with the lower melting temperature of elemental Ga (303 K) as compared to Al (933 K) and the wide liquid range of Ga. Finally, Haberland has investigated the temperature-dependent optical spectra of small Na clusters ($N = 4, 7, 11$) and observed that, while at low temperatures ($T \approx 35$ K) discrete lines are observed (corresponding to transitions between electronic states of solid-like clusters), at high temperatures ($T > 380$ K) broad surface plasmon-type peaks are observed, which are consistent with liquid-like clusters exhibiting collective electronic excitations, as discussed in Section 4.2.3.

There is continuing debate as to whether electronic shell closings actually require the cluster to be liquid-like. A number of calculations have shown that, due to their high symmetry, solid icosahedral sodium clusters exhibit electronic shell structure, which is similar to jellium predictions, up to 1,000 atoms. Other calculations have revealed that the structure of the cluster surface (its symmetry, roughness and facetting) determines the details of the cluster's electronic structure. The states near the Fermi level, which are responsible for electronic control of cluster stability, have their maximum amplitude at the surface and are, therefore, most strongly affected by the nature and structure of the surface.

In general, however, it can be said that, while electronic shell structure is shown by hot, liquid-like clusters, geometric shell structure is exhibited by colder, solid-like clusters. The transition from electronic to geometric shell structures can therefore be achieved by lowering the cluster temperature or increasing the cluster nuclearity.

4.6 THE TRANSITION TO BULK STRUCTURE

In the size regime where geometric shell effects determine cluster structure and stability (i.e. where clusters are solid-like), a variety of cluster geometries are observed (or inferred from experiment). As mentioned above, many clusters of metallic elements are found to adopt icosahedral or decahedral geometries. The structures, which possess five-fold symmetry axes, are incompatible with translational symmetry and therefore involve non-crystalline packing of atoms. Since bulk metals are crystalline, generally adopting fcc, hcp or bcc packing, elements adopting non-crystalline icosahedral or decahedral cluster geometries must undergo a structural phase transition as a function of increasing cluster size. This leads to the important question: 'At what size do metal clusters adopt the structure of the bulk metal?'

4.6.1 Crystalline vs. Icosahedral/Decahedral Packing

There is, as yet, very little direct experimental evidence identifying a critical nuclearity for such a structural phase transition in the gas phase. According to MS experiments, calcium clusters remain icosahedral (rather than adopting the fcc structure of bulk Ca) up to at least 5,000 atoms, while sodium clusters are icosahedral up to tens of thousands of atoms. Aluminium clusters, by contrast, adopt octahedral geometries, based on fcc packing (as in the bulk) for clusters as small as hundreds of atoms. Similar results have been observed for indium. As mentioned earlier, cobalt and nickel clusters, of tens or hundreds of atoms have icosahedral shell structures.

In most calculations of cluster stability, the predicted lowest energy structures are icosahedral for small shell clusters. Small clusters have high surface/bulk ratios and therefore unfavourable surface energies. The surface energy of a cluster can, however, be reduced by adopting non-crystalline, highly coordinated icosahedral or decahedral geometries—as discussed previously (Section 2.2.5) for rare gas clusters. Inspection of Figure 4.13 reveals that, while fcc-like cuboctahedral clusters have eight close-packed (111)-like faces (which are equivalent to the (111) surfaces of an fcc crystal) and six non-close-packed (100)-like faces, icosahedral clusters have twenty pseudo-close-packed (111)-like faces. Non-crystalline icosahedral and decahedral geometries are therefore preferred for small and medium sized clusters (with a high proportion of surface atoms) as these structures minimize surface energy. As the clusters get larger, the elastic strain (a bulk phenomenon—see Section 2.2.5) gets larger. The balance between the two effects therefore depends on the ratio of surface to bulk atoms, which varies as $N^{-1/3}$. Thus, at a certain critical nuclearity, there must be a transition from non-crystalline (low surface energy, high strain) to bulk-like crystalline (strain-free, higher surface energy) structures.

4.6.2 Structure Determination of Supported Clusters

Much of the information about the structures of metal clusters in free cluster beams is obtained indirectly, generally by analysis and modelling of mass spectroscopic data and cluster ion mobility studies. In recent years, however, there has been a significant growth in the application of X-ray diffraction and electron microscopy to the study of cluster structure. These studies have generally been performed on passivated clusters (with organic molecules bonded to the surface of the cluster—see Section 8.2) that have been deposited on solid substrates (*supports*). It is believed that the interaction between the cluster and the passivating ligands and/or the solid support can lead to rearrangment (reconstruction) of the cluster. Depending on the experimental conditions, the deposited clusters may also be metastable, non-equilibrium structures. Care must therefore be taken in correlating structures from diffraction/microscopy measurements on supported passivated clusters with the results of mass spectroscopic (or other) experiments on free beams of naked clusters.

4.6.2.1 Electron Microscopy

TEM and STM studies of surface-passivated (colloidal) metal particles (see Section 8.2) have shown that platinum and palladium colloids grow as fcc single crystals, even for particles with diameters as small as 4 nm. Gold and silver colloids, by contrast, are often found to consist of icosahedral and (Marks) decahedral particles.

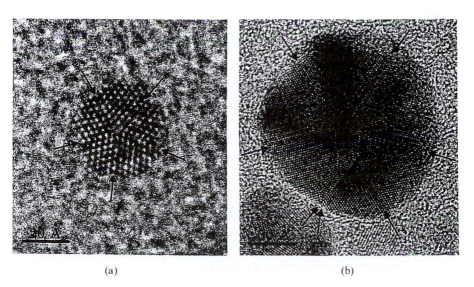

(a)   (b)

Figure 4.15 TEM images of silver particles. (a) 4 nm diameter Marks decahedron. (b) 20 nm diameter multi-twinned fcc structure.

The Marks decahedron (see Figure 4.15a) is formed from the Ino decahedron (Figure 4.13) by introducing (111) facets and reducing the size of the (100) faces, thereby lowering the surface energy of the cluster. Figure 4.15 shows, however, that, while the smaller silver particles ($D \approx 4$ nm, $N \approx 1000$ atoms) are true Marks decahedra (with non-fcc local packing), the larger particles ($D \approx 20$ nm) consist of fcc-packed regions which are multiply twinned into an overall decahedral structure (Figure 4.15b). This is consistent with the arguments put forward above. The formation of defects and dislocations (which relieve strain) enables the formation of local regions of fcc packing, which leads to the observed multiply-twinned structure, rather than a single microcrystal. However, electron microscopy studies of supported clusters which have been heated have shown that multiply twinned structures can be annealed into single domain structures.

4.6.2.2 X-ray Diffraction

In the mid 1990s, Landman, Whetten and co-workers reported a combined experimental and theoretical study of the structures of gold clusters ($N = 100$–1,000, $D \approx 1.4$–3.0 nm) passivated by alkylthiol ligands (see Section 8.2). Comparison was made between the experimental synchrotron X-ray diffraction patterns (from powder and thin film cluster samples) and the patterns calculated for predicted low energy structures, as shown in Figure 4.16. This analysis indicated an abundance of decahedral and fcc-like truncated octahedral clusters, with a transition from decahedral to truncated octahedral occurring at diameters of around 1.7–2.0 nm ($N \approx 200$ atoms).

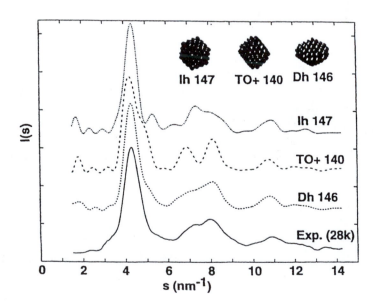

Figure 4.16 Comparison of theoretical and experimental X-ray diffraction patterns for passivated gold clusters, showing decahedral (Dh) and (fcc) truncated octahedral (TO) geometries.

The observed mean lattice contractions are significantly smaller than predicted (for bare clusters), indicating that the surfactant (passivating) ligands lower the energy of the cluster, thereby reducing its tendency for surface relaxation. The enhanced strength of binding of the thiols to the more open (100)-type surfaces of the truncated octahedron help to stabilize this structure relative to other geometries.

In 1999, Lee and co-workers reported X-ray diffraction studies of bare tungsten clusters generated by laser and thermal decomposition of $W(CO)_6$ vapour. Cluster sizes were varied by changing the operating conditions. In addition to tungsten clusters (with greater than 95% purity), some hexagonal tungsten carbide (W_2C) particles were also generated in some runs. Cluster diameters (D) were estimated based on a Scherrer-type analysis of the diffraction peak widths. The broad, featureless patterns observed for small clusters were attributed to amorphous (non-crystalline) W_N clusters. For larger clusters, peaks (which get sharper with increasing cluster size) due to fcc W were identified. For larger particles, additional peaks start to appear which correspond to those observed for bulk bcc W.

The number of atoms (N) in a cluster of diameter D can be estimated as the ratio of the cluster volume (V_c) to effective atomic volume (V_a^*) using an extension of the Spherical Cluster Approximation (Section 1.2.1):

$$N \cong \frac{V_c}{V_a^*} = \left(\frac{D}{2R_0} \right)^3 f \qquad (4.31)$$

where R_0 is the atomic radius (0.148 nm for W) and f is the packing fraction (0.74 for fcc and 0.68 for bcc structures, respectively). Using this formula, Lee *et al.* estimated the average nuclearities of clusters where structural changes were observed. Thus, fcc peaks first appear for clusters of average diameter $\langle D \rangle \approx 3$ nm (corresponding to $\langle N \rangle \approx 900$ atoms), and bcc peaks appear at $\langle D \rangle \approx 7$ nm ($\langle N \rangle \approx 10,500$).

To summarize these experimental results, two size-dependent structure changes are observed: amorphous (possibly icosahedral or decahedral-based) to fcc packing at around 900 atoms; and fcc $\rightarrow$ bcc (bulk-like) at around 10,500 atoms. In a previous theoretical study, Tománek *et al.* predicted the fcc $\rightarrow$ bcc transition to occur at $N = 5,660$, which is in reasonable agreement with these experimental results. The occurrence of the intermediate fcc phase has been rationalized in terms of the fcc-type clusters having more compact structures (since $f_{fcc} > f_{bcc}$) and therefore (for a given value of N) reduced surface areas ($A_{fcc}(N)/A_{bcc}(N) \approx (f_{bcc}/f_{fcc})^{2/3} \approx 0.95$). As the clusters get larger, and the surface/bulk ratio decreases, however, the enhanced electronic (band filling) stability of the bcc structure wins out and bcc-like clusters are formed.

4.7 EXERCISES

4.1   Describe how Liquid Drop behaviour manifests itself in the variation of the properties of metal clusters with their size. Indicate when, and for what reasons, deviations from Liquid Drop behaviour occur.

4.2   Use the jellium energy level diagram shown below to predict the magic numbers that you would expect to see (in the region $N = 2$–50) for the following types of metal clusters:

   (i) uncharged alkali metal clusters, M_N

   (ii) singly charged cationic copper clusters, Cu_N^+

   (iii) singly charged anionic aluminium clusters, Al_N^-.

Energy

$$
\begin{array}{ll}
\underline{\hspace{3cm}} & 1g \\
\underline{\hspace{3cm}} & 2p \\
\underline{\hspace{3cm}} & 1f \\
\underline{\hspace{3cm}} & 2s \\
\underline{\hspace{3cm}} & 1d \\
\underline{\hspace{3cm}} & 1p \\
\underline{\hspace{3cm}} & 1s \\
\end{array}
$$

4.3   Explain why the cationic sodium clusters Na_{19}^+ and Na_{21}^+ have pseudo-spherical geometries, while Na_{17}^+ and Na_{23}^+ are ellipsoidally distorted. Predict the nature (prolate or oblate) of the distortions in the cases of Na_{17}^+ and Na_{23}^+.

4.4   Describe and explain the changes that occur in the mass spectra of sodium clusters as a function of increasing cluster size (in the range 2–25,000 atoms).

4.8 FURTHER READING

Bjørnholm, S., 1994, Shell structure in atoms, nuclei and in metal clusters. In *Clusters of Atoms and Molecules*, Vol.I, edited by Haberland, H., (Berlin, Springer-Verlag), pp. 141–162.

Bonačić-Koutecký, V., Fantucci, P. and Koutecký, J., 1994, Quantum chemistry of clusters. In *Clusters of Atoms and Molecules*, Vol.I, edited by Haberland, H., (Berlin: Springer), pp. 15–49.

Brack, M., 1993, The physics of simple metal clusters: self-consistent jellium model and semiclassical approaches, *Reviews of Modern Physics*, **65**, pp. 677–732.

Broglia, R.A., 1994, The colour of metal clusters and of atomic nuclei. *Contemporary Physics*, **35**, pp. 95–104.

Hall, B.D., Hyslop, M., Wurl and Brown, S.A., 2001, Electron diffraction from atomic cluster beams. In *Fundamentals of Gas-Phase Nanotechnology*, edited by Kish, L., Granqvist, C.G., Marlow, W. and Siegel, R.W., (Dordrecht, Kluwer), pp. 1–32.

de Heer, W.A., 1993, The physics of simple metal clusters: experimental aspects and simple models. *Reviews of Modern Physics*, **65**, pp. 611–676.

Johnston, R.L., 1998, The development of metallic bonding in clusters. *Philosophical Transactions of the Royal Society of London*, **A356**, pp. 211–230.

Kappes, M.M., 1988, Experimental studies of gas-phase main-group metal clusters. *Chemical Reviews*, **88**, pp. 369–389.

Knickelbein, M.B., 1999, The spectroscopy and photophysics of isolated transition metal clusters. *Philosophical Magazine*, **B79**, pp. 1379–1400.

Martin, T.P., 1996, Shells of atoms. *Physics Reports*, **273**, pp. 199–241.

Phillips, J.C., 1986, Chemical bonding, kinetics, and the approach to equilibrium structures of simple metallic, molecular and network microclusters. *Chemical Reviews*, **86**, pp. 619–634.

Whetten, R.L. and Schriver, K.E., 1989, Atomic clusters in the gas phase. In *Gas Phase Inorganic Chemistry*, edited by Russell, D.H., (New York, Plenum), pp. 193–226.

CHAPTER FIVE

Metal Clusters II: Properties

5.1 INTRODUCTION

In the previous chapter, a number of models were presented which have been developed to rationalise experimental data obtained for metal clusters and to describe the variation of their properties with increasing size. In this chapter, I shall describe, in more detail, three particular properties of metal clusters, the experiments that have been developed to study these properties and theories that have been propounded to explain them. The three properties in question are: the reactivity; metal/insulating character; and magnetic properties of metal clusters. Another area of metal cluster research, namely colloidal metal particles, will be discussed in the final chapter (see Section 8.2).

5.2 REACTIVITY OF METAL CLUSTERS

5.2.1 Background

Most studies of cluster reactivity have been performed on transition metals, those elements in the middle of the periodic table which have incompletely filled d orbitals. The interest in transition metal cluster reactivity stems from the desire to model and gain understanding of fundamental processes which occur in heterogeneous catalysis, where the catalysts are typically pure transition metals or intermetallic compounds and the catalytic activity can be controlled by varying the composition of the catalyst. The surface of a metallic catalyst generally lowers the activation barrier to bond breaking and stabilizes reaction intermediates which eventually proceed to form reaction products. In typical industrial catalytic processes, reactor gases are passed through a bed containing finely divided metal particles. In so-called 'structure insensitive' reactions, the activity of the catalyst depends only on total catalyst surface area, so smaller particles (with greater total surface area per given mass of catalyst) show greater activity. In 'structure sensitive' reactions, the dependence of catalytic activity on particle size is more complex, with certain critical sizes being responsible for catalysis.

Most studies of processes relevant to catalysis have concentrated on the reactivity of clean single-crystal metal surfaces under vacuum—a situation very far removed from experimentral conditions. By studying the reactions of clusters with molecules in an inert carrier gas, the conditions of heterogeneous catalysis can be more closely approximated.

It is known that for the surfaces of bulk metals, chemical reactivity is dependent on the number and arrangement of the d electrons and, hence, on the identity of the element and the topography of the surface in question. In the case of finite clusters, however, reactivity also depends on the number of atoms in the cluster and the overall cluster geometry. As will be discussed below, quite dramatic size dependence is observed in cluster reactivity.

In this section, cluster reactivity will be split into first the adsorption of molecules onto clusters and second the reactions of molecules on clusters. Before turning to specific reactions however, I will present a brief overview of experiments which have been designed to study the reactivity of clusters.

5.2.2 Experimental Details

All cluster reactivity experiments consist of three stages: cluster generation; reaction; and detection. The cluster source is usually a laser vaporization source, due to the high refractivity of transition metal elements. Reaction occurs in a device wherein the clusters are in intimate contact with a reactive gas, which is usually present as a mixture with an inert gas. Detection is usually by mass spectrometery—either of ions generated prior to reaction, or of ionized reaction products. Laser ionization accompanied by TOF mass analysis is common. As details of cluster sources and detectors have been presented in Chapter 1, here I will only make a brief mention of the types of apparatus used for studying the reactions of clusters with gas molecules.

5.2.2.1 Fast Flow Reactors

The fast flow reactor is directly attached to the laser vaporization cluster source, with the reactive molecules added to the inert buffer gas (usually helium), so that reaction takes place prior to supersonic expansion (and cooling) of the cluster molecular beam into vacuum.

5.2.2.2 Flow Tube Reactors

In the flow tube reactor, the cluster beam is allowed to expand from the source region into a wider tube in which there is a flow of buffer (typically 3 kPa of He) and reactive gases. The beam is then expanded into vacuum. In the flow tube reactor, there is a better separation of the cluster growth and reaction regions than in the fast flow reactor.

5.2.2.3 Collision Cells

An alternative procedure is to expand the cluster beam and pass it through a reaction cell containing either a pure reactive gas or one that is diluted in an inert gas.

5.2.2.4 Crossed Beam Experiments

By analogy with reactive scattering experiments of small molecules, some cluster reactivity measurements have been obtained by intersecting molecular beams of clusters and reactive molecules and detecting the products as a function of the scattering angle.

5.2.2.5 Experiments with Cluster Ions

In order to investigate the reactivity of size-selected clusters, the initially generated clusters (if neutral) must be ionized and mass selected before reaction. The cluster ions can then be passed through a collision cell, as described above, or they can be injected into an ion trap within which they are exposed to a pure or dilute reactive gas.

5.2.2.6 Comparison of Reactivity Experiments

By varying the conditions of cluster generation and the type of reactivity apparatus, different aspects of cluster reactivity may be probed. Mass selection of ionic clusters enables better control of experiments and gives information on the reactivity of a particular cluster. The disadvantage of mass selection is the reduction in cluster beam intensity, though this can be overcome in ion trap experiments.

In fast flow and flow tube reactors there are a large number of collisions between clusters and gas molecules (inert and reactive), leading to thermalization of the clusters and a narrow distribution of cluster temperatures. The major disadvantage of these methods is the lack of cluster-size selection, so that any kinetic analysis corresponds to the high-pressure limit.

Injection of pre-selected ions into a collision cell or ion trap enables a variety of experimental conditions to be sampled. In the presence of a buffer gas, the clusters are heated by collisions and may be thermalized. In the absence of a buffer gas, the clusters have constant energy. Varying the accelerating potential of the cluster ions allows the investigation of activation barriers to reaction. In the ion trap, multiple collisions are possible and the timescale of the reactivity experiment can extend up to several seconds. This should be contrasted with the other experiments, where the timescale in which reaction can take place ranges from micro- to milliseconds.

Experiments (with ions or neutral clusters) in crossed beams generally result in few collisions (perhaps only one). In the case of mass-selected cluster ions, individual reaction cross sections can be obtained from crossed beam experiments.

5.2.3 Adsorption of Small Molecules

The simplest reaction between a cluster and a small molecule is adsorption, in which the molecule remains bound to the cluster surface, rather then being incorporated into the cluster or reacting with other molecules or molecular fragments. There is obviously a close analogy between adsorption on clusters and that on macroscopic surfaces. At very low temperatures, molecules can be *physisorbed* on the cluster (via weak van der Waals interactions) for short time periods. On bulk surfaces, enthalpies of physisorption (exothermic) are typically less than 50 kJ mol^{-1}. There is generally a small energy barrier to overcome for a physisorbed molecule to become chemisorbed, where the molecule now is covalently bonded to the surface. Typical enthalpies of chemisorption (exothermic) are of the order of 200 kJ mol^{-1}, or even higher. As surface reaction occurs via the chemisorbed state, in this section, I will only consider chemisorption of small molecules on metal clusters.

Considering the chemisorption of the generic molecule *AB* on the metal cluster M_N, two distinct situations are possible:

(a) Non-dissociative adsorption—where the *AB* molecule remains intact:

$$M_N + AB \rightarrow M_N(AB)$$

The process of adsorbing a molecule onto the surface of a metal cluster generally leads to heating (due to the exothermicity of the adsorption process), which may in turn lead to dissociation of the molecule (*vide infra*). The likelihood of non-dissociative adsorption can be increased by coating the metal cluster with rare gas atoms, which evaporate off when the molecule adsorbs, resulting in a relatively cold cluster–molecule complex, for example:

$$Rh_N^+ Ar_m + CH_4 \rightarrow Rh_N^+ (CH_4) + mAr$$

This technique, which is analogous to the soft-landing technique used for low energy deposition of clusters on bulk surfaces (see Section 8.4), can be used to isolate relatively unstable species, stabilize reaction intermediates and even to alter reaction pathways and generate new products.

(b) Dissociative adsorption—where the *AB* molecule breaks up into atoms *A* and *B* which are bound to the surface of the cluster:

$$M_N + AB \rightarrow M_N(A)(B)$$

For polyatomic molecules, more complex dissociative adsorption pathways are possible, such as:

$$M_N + AB_3 \rightarrow M_N(AB_2)(B) + M_N(AB)(B)_2 + M_N(A)(B)_3...$$

The nature of the adsorption products depend on the activation barriers to bond breaking, the energies of the cluster and reactant molecule, the *A–B* bond strength and the strength of binding between the molecule (and molecular fragments) and the surface of the cluster.

5.2.3.1 Adsorption of Ammonia on Transition Metal Clusters

Some of the earliest cluster reactivity experiments were performed by Riley and Parks in the mid-1980s. Using a flow tube reactor, they measured the number of molecules adsorbed by clusters as a function of the partial pressure of reactant gas. One of the reactions that they studied was the non-dissociative adsorption of ammonia on iron, cobalt and nickel clusters:

$$M_N + mNH_3 \rightarrow M_N(NH_3)_m$$

Figure 5.1 shows a plot of the average number (m) of ammonia molecules adsorbed on the Fe_{61} cluster, as a function of the partial pressure of NH_3. For $p(NH_3) < 0.2$ Pa, the reaction is kinetically controlled and the uptake (m) is time dependent, thus the longer the clusters are exposed to ammonia, the more ammonia is adsorbed. For $p(NH_3) > 0.2$ Pa, the reaction is in equilibrium (thermodynamically controlled) and the uptake is independent of time.

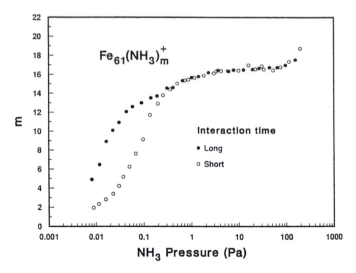

Figure 5.1 The average number (m) of NH_3 molecules bound to Fe_{61} as a function of the partial pressure of NH_3.

The reason for the coverage-induced switch from kinetic to thermodynamic control is that the binding energy of NH_3 to the Fe surface decreases with increasing coverage. At low coverage the binding energy is large enough that the time for NH_3 desorption is greater than the time that the clusters spend in the flow reactor. In this case, the number of molecules adsorbed depends on the number of cluster-molecule collisions—which depends on exposure time for fixed $p(NH_3)$. At higher coverages, the binding energy becomes small so the lifetime for NH_3 desorption becomes less than the time spent in the flow reactor and an equilibrium situation exists—with molecules continually adsorbing and desorbing from the cluster. At higher partial pressures, the coverage (uptake) flattens off, becoming independent of $p(NH_3)$. Such a plateau indicates that the sites of higher binding energy have been filled and the next available sites have significantly lower binding energies. The cluster can be said to be 'saturated'. The number of molecules corresponding to saturation indicates the number of strong binding sites, which can in turn provide information on the cluster structure. In the case of Fe_{61}, the plateau corresponds to the adsorption of sixteen ammonia molecules.

Figure 5.1 shows that there is an increase in m for $p(NH_3) > 100$ Pa. This high-pressure increase in ammonia uptake is characteristic of the onset of molecular physisorption—i.e. the start of a second layer of ammonia molecules which are much more weakly bound to the cluster.

Small molecules can be used as chemical probes of cluster structure by studying the extent of adsorption as a function of pressure and cluster size. A number of geometries that have been proposed for small transition metal clusters are shown in Figure 5.2.

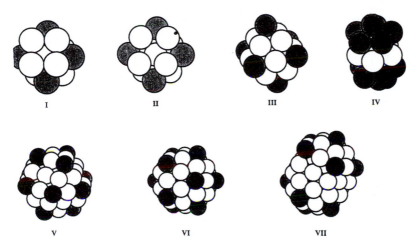

Figure 5.2 Structures of metal clusters, postulated on the basis of NH_3 uptake experiments on iron, cobalt and nickel clusters. Primary binding sites are shaded.

Ammonia uptake experiments by Parks and Riley on Fe_{13} revealed two distinct plateaus, one with $m = 4$ and another (at higher pressure) with $m = 6$. These results have been interpreted in terms of four equivalent strong binding sites and two more weakly binding sites. As surface studies and theoretical calculations indicate that the NH_3 molecule binds to metal surfaces via donation of the N-localized lone pair to a single metal atom and, assuming that the same bonding pattern holds for cluster–NH_3 binding, Parks and Riley have suggested that either the fcc-like cuboctahedral cluster (structure 1 in Figure 5.2) or the incomplete bcc-like rhombic dodecahedral cluster (II) are consistent with these results. Similarly, the plateau at $m = 6$ (with onset at $p(NH_3) = 0.05$ Pa) observed for Co_{19} has been attributed to the fcc-like octahedral geometry (III). By contrast, the Ni_{19} cluster saturates at $m = 12$, though the saturation pressure is much higher (13 Pa)—which has been attributed to this cluster having the double icosahedral geometry (IV). The closer proximity of the twelve binding sites in (IV) would appear to explain the relatively high pressure required for saturation. Saturation at $m = 12$ is common for M_N ($M =$ Fe, Co, Ni) in the range $N = 19$–34. Both Co_{55} and Ni_{55} have extensive plateaux at $m = 12$, which is consistent with either fcc-like cuboctahedral (V) or non-crystalline icosahedral (VI) geometries, both structures having twelve isolated and exposed vertices—which are expected to be strong binding sites for NH_3 adsorption. There are no closed geometric shell fcc-like clusters for 71 metal atoms, so the marked plateaux (at $m = 12$) in ammonia adsorption measured for Co_{71} and Ni_{71} have been interpreted in terms of the 2-shell double icosahedral structure (VII), which has twelve exposed vertices. The results of Parks and Riley's experiments for M_{50}–M_{160} ($M =$ Co, Ni) indicate that icosahedral growth predominates in this size regime—which is in agreement with the mass spectral measurements mentioned previously (see Section 4.5.3.3).

5.2.3.2 Thermodynamics of Ammonia Adsorption

Under equilibrium conditions, the addition of the mth adsorbate molecule (A) to a cluster (M_N) can be represented as:

$$M_N A_{m-1} + A \rightleftharpoons M_N A_m$$

with the equilibrium constant given by:

$$K_{eq} = \frac{\left[M_N A_m \right]}{\left[M_N A_{m-1} \right] \left[A \right]} \tag{5.1}$$

Elementary thermodynamics relates the equilibrium constant to the standard Gibbs free energy of reaction: $\Delta G^0 = -RT\ln(K_{eq})$ and, hence, to the standard enthalpy (ΔH^0) and entropy (ΔS^0) of reaction (via $\Delta G^0 = \Delta H^0 - T\Delta S^0$). K_{eq} can be calculated by assuming that relative cluster concentrations are proportional to their relative mass spectral intensities. Assuming that ΔH^0 and ΔS^0 are approximately constant over a small temperature range, allows them to be obtained from the gradient and intercept, respectively, of a van't Hoff plot of $\ln(K_{eq})$ against $1/T$. Such plots are shown in Figure 5.3 for the adsorption of the twelfth ammonia molecule on Co_{55} and Co_{71} clusters. As the binding energy of an NH_3 molecule is effectively given by $-\Delta H^0$, it can be seen that the twelfth ammonia molecule is more strongly bound to Co_{71} ($\Delta H^0 = -68.6$ kJ mol^{-1}) than to Co_{55} ($\Delta H^0 = -61.1$ kJ mol^{-1}). This may be because the fractional coverage of $Co_{71}(NH_3)_{12}$ is less than that of $Co_{55}(NH_3)_{12}$.

5.2.3.3 Reactions of Transition Metal Clusters with Hydrogen

Hydrogen and deuterium molecules adsorb dissociatively on transition metal surfaces and this reactivity is also observed on transition metal clusters:

$$M_N + xH_2 \rightarrow M_N H_{2x}$$

In 1985, Smalley and co-workers showed, using a flow tube reactor, that the dissociative adsorption of D_2 on Nb_N exhibits strong size-dependence, with some clusters ($N = 8$, 10, 16) being completely unreactive, while others react readily. The low reactivity of certain nuclearities was found to correlate with high ionization energies of the corresponding neutral Nb clusters. Such correlations between reactivity and ionization energies—a drop in IP from M_N to M_{N+1} is often accompanied by a decrease in reactivity towards hydrogen—have been observed by the groups of Whetten and Riley, for Fe and other transition metals. A plausible explanation for this observation is that clusters with lower ionization potentials can more readily donate electrons into the antibonding (σ^*) orbital of H_2, thereby facilitating dissociative chemisorption. The agreement is not complete, however, and other correlations (such as with the cluster HOMO–LUMO gap) have been suggested. In some cases, Smalley also observed that neutral and charged clusters exhibit very similar reactivities, which may indicate that, in these cases, geometric (cluster shape), rather than electronic factors are dominant. In the Phillips model of cluster reactivity, structural factors (cluster topology and local surface geometry) are vital in determining cluster reactivity, while electronic factors, such as cluster IPs and magnetism, play secondary roles. As will be discussed below, however, it is difficult to disentangle electronic and structural effects as the electronic structure of a cluster depends critically on its geometry.

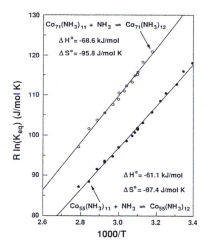

Figure 5.3 van't Hoff plots for the addition of a twelfth NH_3 molecule to $Co_{55}(NH_3)_{11}$ (filled circles) and $Co_{71}(NH_3)_{11}$ (open circles). The straight lines are linear least squares fits to the data, with the resulting ΔH^0 and ΔS^0 values indicated.

In the reaction of Nb_{12} (both neutral and with a single positive charge) with D_2, flow tube reactor experiments have shown that half of the clusters react rapidly, with the remaining half only reacting at high D_2 pressures. This has been interpreted in terms of two distinct isomers (with roughly equal abundance) with very different reactivities towards D_2 adsorption.

Measurements have been made of the variation of the compositions of hydrogen saturated metal clusters, with increasing cluster size. The saturation compositions have been compared with the prediction of the Structureless Packing Model (SPM)—in which the cluster is treated as spherical, with the thickness of the cluster surface layer being equal to the cube root of the atomic volume. In the structureless packing model, the ratio of surface atoms (N_s) to the total number of atoms (N) in a cluster is approximately given by:

$$\frac{N_s}{N} \approx 1 - \left[1 - \left(\frac{4\pi}{3N} \right)^{1/3} \right]^3 \tag{5.2}$$

In the case of nickel clusters (with 40–260 atoms), the measurements indicate good agreement with the SPM, with approximately one hydrogen (or deuterium) atom per surface Ni atom. Iron clusters, however, tend to show lower saturation coverages than nickel clusters, with the fractional surface coverage decreasing with increasing size—dropping from 1.0 for Fe_{40} to 0.8 for Fe_{250}. These results have been used to explain the different products produced in the Fischer–Tropsch synthesis of hydrocarbons from CO and H_2, when using Ni or Fe particles as catalysts. In the case of Ni clusters, there is a high likelyhood that the adsorbed CO molecules will be surrounded by H atoms, due to the high saturation coverage of H on the Ni surface. In this case, the major Fischer–Tropsch product is CH_4. For iron, however, the lower surface coverage by H allows CO molecules to adsorb to adjacent sites, permitting the formation of C–C bonds and, hence, the generation of higher hydrocarbons.

When hydrogen adsorbs dissociatively to a metal (or cluster) surface, the H atoms bind to two, three or four metal atoms in a bridging mode, rather than being bound to a single metal atom in the on-top (terminal) mode—as for NH_3. The different coordination sites of H and NH_3 is confirmed by the fact that hydrogenated metal clusters can adsorb as many ammonia molecules as unhydrogenated clusters—i.e. H and NH_3 bind non-competitively. While H-adsorption does not block sites for NH_3 adsorption, it does reduce the strength of M–NH_3 bonding and tends to emphasize the relative stability of adsorption of twelve ammonia molecules.

The strength of cluster-adsorbate interactions has been estimated from IP measurements and IR (photodepletion) spectra of cluster-molecule complexes. Knickelbein and co-workers have used these techniques to prove that the adsorption of H_2 (or D_2) on metal clusters is accompanied by molecular dissociation, which is in agreement with theoretical simulations by Jellinek.

5.2.3.4 Kinetics of Hydrogen Adsorption

The dissociative adsorption of a hydrogen molecule on a metal cluster:

$$M_N + H_2 \rightarrow M_N H_2$$

obeys the second order rate equation:

$$\frac{d[M_N]}{dt} = -k[M_N][H_2] \tag{5.3}$$

If the concentration of hydrogen [H_2] is large and constant, and assuming that the above reaction is the only one that can occur (i.e. in the single collision regime, with irreversible adsorption) Equation (5.3) reduces to a pseudo-first order rate expression, which integrates to give:

$$\ln[M_N] - \ln[M_N]_0 = -k\Delta t[\text{H}_2] \tag{5.4}$$

where Δt is the interaction time, which is known from the geometry of the flow tube reactor and the experimental conditions, and the metal cluster concentration $[M_N]$ is assumed to be proportional to its mass spectral intensity. The pseudo-first order rate constant k is then obtained as the slope of a plot of $\ln[\text{Fe}_N]$ against $[\text{H}_2]$.

As shown in Figure 5.4, for the reaction of Fe clusters with H_2, the reactivity (as measured by the rate constant, k) can vary by as much as 3 orders of magnitude with cluster size. Similar results have been observed for the reaction of hydrogen with other metals, such as Co, V and Nb, though smaller size-dependent variations are observed for Ni clusters. Interestingly, while the reaction of nitrogen (N_2) with Nb and Co clusters shows similar variation to hydrogen, the isoelectronic carbon monoxide (CO) molecule displays a variation of at most two to three in reactions with clusters of a variety of transition metal elements. Small size variations are also observed for O_2 adsorption.

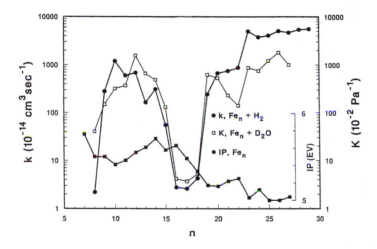

Figure 5.4 Dependence on iron cluster size of reaction rate constants with H_2 (k, left scale, filled circles), equilibrium constants for binding D_2O (K, right scale, open circles), and ionization potentials (IP, inset scale, filled squares).

5.2.3.5 Comparison of Thermodynamics and Kinetics

As discussed above for the reaction of hydrogen with iron clusters, in kinetically controlled reactions, the reaction rate constant (k) often exhibits dramatic size-dependence. It has been demonstrated that this variation can roughly be correlated with the ionization energy of the cluster. In thermodynamically controlled reactions, such as the high pressure adsorption of ammonia on iron clusters, the equilibrium constant (K_{eq}) may also exhibit large size-dependent variations.

Considering the adsorption of D_2O on iron clusters, the relatively weak binding of water (or heavy water) to metal surfaces means that the reaction is thermodynamically controlled, even at low coverage. It is remarkable that a plot of K_{eq} against N for the D_2O/Fe_N reaction follows almost exactly that of k against N for the H_2/Fe_N reaction. As Riley has stated, the fact that thermodynamic and kinetic parameters show the same dependence on N indicates that there must be a fundamental property of the cluster that governs its chemistry, as well as its physical properties (such as the IP). Riley has concluded that this underlying factor is the geometrical structure of the cluster and that the structures of iron clusters undergo many changes in the range $N = 8$–25, so that the reactions of both H_2 and D_2O with Fe clusters are structure sensitive.

The link between thermodynamics and kinetics has been strengthened by the results of photodissociation studies of N_2 and CO from charged Nb clusters:

$$Nb_N(XY)^+ \ (XY = N_2, CO) + h\nu \rightarrow Nb_N^+ + XY$$

demonstrate that the most reactive clusters form the strongest cluster–XY bonds.

In general, for transition metal clusters, high IP and low reactivity are characteristics of stable clusters and that kinetic and thermodynamic reactivity may be related, as electronic and geometric structures are intimately related. This is consistent with flow tube reactor studies of the reaction of aluminium, copper and gold clusters, where closed jellium electron shell clusters were found to be significantly less reactive towards reaction with O_2 and H_2O than open shell clusters. Similarly the greater stability of even over odd electron clusters, which is manifest in mass spectral abundances and IP data, is reflected in the lower reactivity of clusters with even numbers of electrons.

5.2.3.6 Adsorption of N_2 and CO_2 on Transition Metal Clusters

As mentioned above, the adsorption of nitrogen molecules on transition metal clusters shows similar size-dependence to that observed for hydrogen However, unlike hydrogen, due to the strength of the N≡N triple bond, the nitrogen molecule is not dissociated upon adsorption. As for ammonia, the nitrogen molecule binds to a single metal atom via the lone pair of electrons on one of the nitrogen atoms. Unlike ammonia, however, due to the weaker M···N bonding and the less bulky nature of N_2, a nitrogen molecule can bind to every metal atom on the surface of the cluster—in fact some metals can bind two N_2 molecules. Detailed study of the cluster coverage as a function of $p(N_2)$ can reveal the different binding energies of nitrogen to vertex, edge and face-localized metal atoms, while the saturation coverage gives the number of binding sites (i.e. the number of surface atoms). Thus, Parks and Riley have used N_2 as a chemical probe of the structures of pure Ni_N clusters, as well as mixed Ni_NAl_M clusters (Al atoms do not bind N_2 molecules), with their results being correlated with theoretical studies by DePristo and Jellinek. On the basis of these experiments, clusters in the range Ni_{46-48} are predicted to have fcc or mixed fcc–hcp packing, though at Ni_{49} there is a transition to icosahedral structures, in agreement with mass spectral evidence, magnetic measurements and recent electric polarizability measurements by Knickelbein.

The carbon dioxide molecule (CO_2) is known to bind weakly to bulk transition metal surfaces. Study of the reaction of CO_2 with Nb_N^+ clusters have shown that on very small clusters ($N = 3$–7), CO_2 dissociates:

$$Nb_N^+ + CO_2 \rightarrow Nb_NO^+ + CO$$

while for larger clusters, the CO_2 adsorbs non-dissociatively.

5.2.3.7 Adsorption of CO on Transition Metal Clusters

The carbon monoxide molecule (CO) is isoelectronic with N_2, but it binds much more strongly to transition metals (via a lone pair on the carbon atom). While the nitrogen molecule generally binds to a single metal atom on a metal surface or cluster, CO can bind to one, two, three or four metal atoms, due to the presence of low-lying diffuse π^* antibonding orbitals, which are mainly localized on the carbon atom, and which can accept electron density from the metal atoms (so called 'back donation'). This strength of M–CO bonding is reflected in the rich field of transition metal carbonyl cluster chemistry—with a large number of cluster molecules, $[M_N(CO)_x]$ (usually carrying a negative charge), having been synthesized over the past thirty years. The CO molecules (ligands) in these molecules passivate the clusters (preventing aggregation) so that they are stable in solution and in the solid state.

Flow tube reactor experiments have also been performed to probe the adsorption of CO on preformed naked transition metal clusters. In experiments with niobium and cobalt clusters, Smalley found that, while N_2 showed considerable variation of reactivity with cluster size, the reaction with CO showed a slow monotonic rise in reactivity with increasing cluster size. The reaction with CO is faster than with N_2, which is consistent with the greater strength of M–CO as compared with M–N_2 bonding.

Wöste and co-workers studied the reaction of mass-selected nickel clusters, Ni_N^+ ($N = 2$–20) with CO. Three series of products were detected:

$$Ni_N^+ + mCO \rightarrow [Ni_N(CO)_x]^+ + [Ni_NC(CO)_y]^+ + [Ni_{N-1}(CO)_z]^+$$

Wöste noticed that, in most cases, the maximum number of CO molecules that could be adsorbed by a particular cluster is such that the resulting metal carbonyl cluster (if neutral) would obey the electron counting rules developed by Lauher, Wade and Mingos for closed shell triangulated geometries. Thus, Ni_4^+ can adsorb a maximum of ten CO molecules, yielding a species $[Ni_4(CO)_{10}]^+$ with 59 valence electrons. The electron counting rules predict that a stable, closed shell tetrahedral metal carbonyl cluster should be characterized by 60 electrons. Similarly, the saturated $[Ni_6(CO)_{13}]^+$ cluster has 85 valence electrons, while the prediction for an octahedral cluster is 86 electrons. This would seem to indicate that the products $[Ni_4(CO)_{10}]^+$ and $[Ni_6(CO)_{13}]^+$ are tetrahedral and octahedral, respectively.

Analogous experiments have been performed on cationic cobalt clusters by Castleman and co-workers. Saturated species, such as $[Co_4(CO)_{12}]^+$ and $[Co_5(CO)_{14}]^+$ have 59 and 72 valence electrons respectively, compared with ideal electron counts of 60 for the tetrahedron and 72 for the trigonal bipyramid. At low pressures, fewer CO molecules are adsorbed, leading to species such as $[Co_5(CO)_8]^+$. Kinetic studies also show that the uptake of eight CO molecules is rapid, while the next six molecules adsorb relatively slowly. This data may be rationalized in terms of the extra steric (ligand–ligand) repulsions which must be overcome in fitting fourteen CO molecules around the Co_5 core.

In the work and Wöste and Castleman, it has generally been assumed that the geometry of the metal cluster does not change when CO is adsorbed. However, the adsorption of CO is an exothermic process ($\Delta H = -E_b$, where E_b is the cluster–CO binding energy) which results in heating of the cluster. It is, therefore, possible that ligand-induced cluster restructuring could occur upon adsorption of m CO molecules, if the total binding energy of m CO ligands is greater than the rearrangement energy of the cluster. For bulk metals, ligand-induced surface reconstruction is a well known phenomenon and theoretical calculations by Jellinek and others have shown that similar effects can also occur for clusters.

Brechignac and co-workers have derived rate constants and binding energies for the adsorption of CO on size-selected nickel clusters, by measuring the photoinduced desorption of CO as a function of CO pressure and laser light intensity. Grushow and Ervin have studied the collision-induced desorption of CO from $[Pt_3(CO)_x]^-$, in collisions with Xe atoms of varying kinetic energy. From these experiments, they have derived cluster–CO binding energies for $x = 1$–6.

5.2.3.8 The Effect of Adsorption on Cluster IP

Separate experiments by Knickelbein and Parks *et al.* have shown that adsorption of ammonia leads to a decrease in the ionization energy of nickel clusters. The shift in IP is particularly large for smaller clusters. The decrease in IP has been explained in terms of dipoles arranged around a metallic sphere. An alternative explanation is that electron donation from the lone pair of electrons on the ammonia molecule to the metal results in a rise in the Fermi energy of the cluster and thus a decrease in its work function or IP. Similar studies by Persson and co-workers on the IPs of copper clusters and their oxidized counterparts, Cu_NO_2, reveal a small increase in IP, presumably due to electron transfer from the metal cluster to the coordinated oxygen.

5.2.3.9 Infrared Spectroscopy of Adsorbates on Clusters

Using a CO_2 infrared laser ($\lambda = 9$–11 μm), Knickelbein and co-workers have measured the IR (vibrational) absorption spectra of a number of molecules adsorbed on clusters. In these experiments, multiphoton absorption by the adsorbed molecule leads to fragmentation and, thus, a depletion in the corresponding peak in the mass spectrum. The depletion (and hence absorption) efficiency is measured as a function of IR frequency, enabling the IR absorption spectrum to be mapped out.

In studies of ethylene (C_2H_4) and ethylene oxide (C_2H_4O) adsorbed on silver clusters, Knickelbein found vibrational frequencies close to those measured for the free molecules in the gas phase. It was concluded that both C_2H_4 and C_2H_4O are molecularly adsorbed on silver clusters, as is known to be the case for adsorption on bulk silver surfaces.

5.2.4 Reactions on Clusters

5.2.4.1 Adsorbate Decomposition

If clusters are to be used as models for catalytic processes taking place on the surfaces of metal particles and solids, then it is necessary to study chemical processes—i.e. bond breaking and formation—on the surfaces of clusters. Most studies to date have involved the adsorption of hydrogen-containing molecules, breaking of one or more of the C–H, N–H or O–H bonds and subsequent desorption of H_2, as one of the reaction products. Examples of this type of study include the dehydrogenation of benzene on Nb clusters and of cyclohexane on Pt clusters.

Although the adsorption of a single water molecule on Fe clusters is a reversible, non-dissociative adsorption process, the addition of a second water molecule results in an irreversible loss of H_2, which desorbs from the cluster. This is believed to be a concerted reaction, involving one H atom from each water molecule, which is driven by the formation of strong cluster–OH bonds:

$$Fe_N^+ + 2H_2O \rightarrow [Fe_N(OH)_2]^+ + H_2$$

In the reaction of ammonia with nickel clusters, short interaction times lead to non-dissociative adsorption of NH_3. For long interaction times, however (especially for high coverages), extensive loss of H_2 is observed, indicating dissociation of some of the ammonia molecules, to generate cluster-bound NH fragments. The amount of H_2 evolved is time dependent, showing that H atoms are bound to the cluster until two collide and recombine:

$$[Ni_N(NH_3)_m]^+ \rightarrow [Ni_N(NH_3)_{(m-p)}(NH)_p(H)_{2q}]^+ + (p-q)H_2$$

This is confirmed by adding D_2 to clusters on which NH_3 has been allowed to interact for a long time. In this case, complex mass spectral signals are observed, indicating that H/D exchange has taken place.

The reaction of ethylene (C_2H_4) with Pt clusters leads to extensive dehydrogenation, with the degree of dehydrogenation increasing with the number of ethylene molecules adsorbed. In contrast to the Ni_N/NH_3 case, however, the number of ethylenes adsorbed is strongly dependent on time, while the amount of H_2 evolved is independent of time and there is no hydrogen left adsorbed to the cluster. This has been interpreted in terms of an equilibrium adsorption of ethylene on the Pt cluster followed by the rate limiting ethylene decomposition step—with every decomposition leading to H_2 loss—presumably via a concerted process:

$$Pt_N^+ + m(C_2H_4) \rightarrow [Pt_N(C_2H_4)_m]^+ \rightarrow [Pt_N C_{2m} H_{(4m-2p)}]^+ + pH_2$$

Michl and co-workers have studied the dehydrogenation of hydrocarbons by Ni clusters. They found that for $Ni_2^+ - Ni_5^+$ the dominant decomposition pathway for butane (C_4H_{10}) involves the loss of two H_2 molecules:

$$\text{Ni}_N^+ + C_4H_{10} \rightarrow [\text{Ni}_N(C_2H_6)]^+ + 2H_2$$

while for $\text{Ni}_6^+ - \text{Ni}_{10}^+$, more extensive dehydrogenation occurs—three molecules of H_2 are lost, to yield $[\text{Ni}_N(C_2H_4)]^+$. Experiments in which Ni_2^+ and Ni_3^+ were collisionally cooled prior to reaction with butane revealed that cooling leads to a greater proportion of reactions which involve the loss of two H_2 molecules—with less cluster fragmentation and alternative reactions. It was, thus, demonstrated that cluster reactivity can be made more specific by carefully controlling the internal energy, as well as the size distribution, of the clusters.

Pd clusters were found to undergo similar reactions with butane to the Ni clusters, though the dehydrogenation activity of Pd clusters exceeds that of Ni: Pd_4^+ removes three H_2 molecules from C_4H_{10}, whereas Ni_4^+ only removes two.

5.2.4.2 Catalytic Formation of Benzene on Iron Clusters

Schnabel and co-workers have studied the reaction of ethylene (C_2H_4) with Fe_N^+ clusters in an ion trap. They observed dehydrogenation of the ethylene molecules to generate surface bound acetylene (C_2H_2) units, which trimerize to form benzene (C_6H_6). In the ion trap, collisions lead to the dissociation of the benzene molecules from the cluster surface.

5.2.5 Reactivity and Catalytic Activity of Supported Clusters

5.2.5.1 Introduction: Clusters in Catalysis

Since the 1960s, research in organometallic chemistry (the chemistry of molecules containing metal–carbon bonds) has led to significant advances in the area of homogeneous catalysis—where reactants, catalyst and products are all in the same phase (usually in solution). Solution-phase organometallic cluster compounds have been studied as homogeneous catalysts, since they can also model some of the aspects of metal surfaces (which are typically used for two-phase heterogenous catalysis). Clusters as homogeneous catalysts, however, suffer from a tendency to decompose, via ligand loss, aggregation (coalescence) and the formation of large colloids or metal particles, which precipitate out of solution.

Many modern commercial heterogeneous catalysts, however, are actually fabricated from small metal particles which are dispersed on a high-surface area non-metallic (usually an oxide) support. In this way, the total surface area available for catalysis, for a given mass of metal catalyst, is increased—leading to greater catalytic activity. Supported metal clusters (nanometre-sized metal particles) are tethered to the support, thereby immobilizing them and stabilizing them with respect to coalescence. Supported clusters also offer the advantage of ease of catalyst recovery and separation from the reactant/product mixture.

5.2.5.2 Supported Clusters

In catalytic systems, support materials are generally oxides, such as silica (SiO_2), alumina (Al_2O_3) or titania (TiO_2). The clusters (metal particles) may be coordinated to the exterior surface of the support, or to the large interior surfaces of nanoporous materials, such as zeolites (alumino silicates with interior tunnels and cages) or nanoporous alumina membranes. These nanoporous materials offer very high surface areas per unit volume and therefore high catalyst dispersion. Supported clusters may be bare, or they may be coated by a shell of ligands. Such ligands may be used to tether the cluster to the support surface: they also serve to prevent cluster coalescence (see the discussion of colloidal metal particles in Section 8.2).

Supported clusters (see Section 8.4) have been extensively studied by a variety of physical techniques, ranging from electronic and vibrational spectroscopy to electron microscopy and X-ray diffraction. These experiments have allowed the determination of the electronic states of supported clusters, and the identities of molecules adsorbed onto them. The technology of cluster deposition from molecular beams (generally onto model surfaces, such as graphite or silicon) is also well developed (see Section 8.5)—so that it is now possible to deposit a fairly *monodisperse* (narrow size-distribution) array of clusters onto the surface of a support. Such techniques enable better control of cluster size than standard solution phase chemical syntheses of clusters. The study of cluster mobility on surfaces, as a function of cluster size, impact energy and substrate temperature, also provides important information on how to avoid cluster migration and growth (coalescence).

There is evidence that the support does not always play a passive rôle when it comes to the reactivity (or catalytic activity) of supported clusters. The so-called 'support interaction' may arise due to the opening up of new cluster relaxation channels, through support-induced changes in the geometric and/or electronic structure of the cluster.

5.2.5.3 Studies of Reactivity and Catalysis by Supported Clusters

Xu and Goodman have investigated the catalysis of the reaction between CO and NO by Pd particles on an oxide support. The products of this reaction are generally CO_2, N_2 and N_2O. In this study, it was found that N_2O was only produced on larger particles (with diameters > 5 nm). Later work has shown that, while larger Pd clusters are more active towards NO reduction, the efficiency of CO oxidation is less dependent on cluster size.

Schmid and co-workers have investigated the effect of ligands on the activity and selectivity of heterogeneous catalysis by metallic clusters. They studied the selective hydrogenation of hex-2-yne to *cis*-hex-2-ene catalysed by Pd clusters (with diameters in the range 3.0–3.6 nm) supported on TiO_2 and active carbon. Compared with bare clusters, covering the clusters with ligands (phenanthrene) results in a decrease in catalytic activity—though this may be compensated for by the increased stability of the catalyst. On changing the ligand from phenanthroline to 3-*n*-decylphenanthrene (a ten carbon atom chain replaces a hydrogen atom on the third carbon atom of the phenanthrene), the activity of the catalyst is greatly reduced—presumably the bulkier ligand blocks more of the active surface—and more side products (such as the *trans* isomer of hex-2-ene and the fully hydrogenated hexane) are formed. When a smaller (*n*-butyl) substituent is used, the rate of production of *cis*-hex-2-ene is slow, but very little of the alternative products are formed—the selectivity of the reaction is greater than that with the unsubstituted phenanthrene ligand.

Porous membranes of Al_2O_3 can be synthesized with a high density of parallel pores or nanotubes. The walls of these pores are coated with OH groups, which can readily be functionalized by reaction with compounds such as $(RO)_3Si(CH_2)_3Y$, where Y is a functional group (e.g. NH_2 or SH) which is chosen depending on the bonding preferences of the metal in question. Gold and palladium clusters have been trapped inside the pores either by sucking a cluster solution through the membrane or by evacuating the membrane then immersing it in the cluster solution. Such systems show promise for future generations of heterogeneous catalysts.

Xu *et al.* have used ligand stabilized carbonyl clusters of iridium ($Ir_4(CO)_{12}$ and $Ir_6(CO)_{16}$) as precursors for Ir hydrogenation catalysts, by absorbing the carbonyls onto MgO or Al_2O_3 and then removing the CO ligands by heating. The Ir_6 cluster was found to be less efficient than Ir_4 and both were less active than microscopic Ir and Pt particles for the hydrogenation of toluene. The catalytic activity of the small clusters is reduced, relative to the larger particles, because they bind H_2 more strongly. Johnson, Thomas and co-workers have investigated catalysis by mono- and bimetallic particles generated by tethering and thermally decomposing iridium (and other metal) carbonyl clusters inside porous zeolite and aluminium phosphate (ALPO) hosts. Such hosts offer large internal surface areas for tethering catalytic particles and, hence, high catalytic activity.

An interesting size-dependence of catalytic properties has recently been observed for gold clusters. Although bulk gold is one of the least catalytically active metals, gold clusters dispersed on thin (2–10 nm) oxide film supports, can catalyse the oxidation of CO to CO_2 at temperatures as low as 40 K! Studies of Au_N clusters supported on TiO_2 have shown that the catalytic activity of the gold clusters increases as the cluster diameter decreases (for $D > 3.5$ nm), with the onset of catalysis being related to the appearance of non metallic properties (see Section 5.3). In the size range $D < 3$ nm, however, the catalytic activity decreases with further decrease in D.

5.2.5.4 Catalysis by Bimetallic Particles

The properties, including catalytic activity, of metals may be modified and fine-tuned by alloying—i.e. forming bimetallic solids. The same is true for small metal particles and clusters and the field of 'nanoalloys' is now attracting a lot of attention. According to Schmid, in the field of catalysis, the mutual influence of different neighbouring atoms can lead to catalytic behaviour which is different (and often better) than that of the monometallic clusters—i.e. synergistic effects are observed. Layered ('core–shell') bimetallic clusters offer fascinating prospects for the design of new catalysts.

Schmid and co-workers have prepared layered colloidal Pd-covered Au clusters (Au/Pd) and Au-covered Pd clusters (Pd/Au) (see Section 8.2 for details). These bimetallic clusters were investigated for their catalytic activity for the hydrogenation of hex-2-yne to *cis*-hex-2-ene, when supported on TiO_2. The presence of the Au core was found to lead to a dramatic increase in catalytic activity compared to pure Pd clusters, with the influence of the (fixed size) core decreasing with increased thickness of the Pd shell. The unique catalytic behaviour of Pd–Au clusters (compared to pure Pd or Au clusters) arises because their electronic structures are quite distinct from those of the pure metals (due to their differing atomic electron configurations and electronegativities).

Finally, a clear synergistic effect has been observed in the catalytic hydrogenation of crotonic acid to butanoic acid by bimetallic $Pt_{20}Rh_{80}$ colloids—with the most active catalysts having a concentration gradient, with increasing Rh concentration towards the surface of the cluster.

5.3 THE METAL–INSULATOR TRANSITION IN CLUSTERS

5.3.1 The Metal Insulator Transition

Part of the continued fascination in metal clusters derives from their position as mesoscopic intermediates between the atomic (non-metallic) and bulk (metallic) regimes. In this Section, I will concentrate on the development of metallic properties with cluster size, and the transition from non-metal-like (or insulator-like) to metal-like behaviour. Particularly interesting, in this respect, is the fundamental question, posed by Edwards and Sienko: 'How many atoms maketh metal?' From common experience, it is obvious that the macroscopic metal is a very good conductor of electricity. But what of the situation in an isolated, finite fragment of metal comprising, say, 10^2–10^4 atoms? And what does electrical conductivity mean in the context of an isolated particle of finite size? Furthermore, how can one begin to probe the property of conductivity in this mesoscopic-size regime?

Within the context of the metal–insulator transition in metal clusters, two situations can be distinguished. The first is the inevitable transition to insulating (non-metallic) behaviour within an individual particle of metal as its dimensions are successively reduced. Throughout this process, the electron wavefunction is assumed to be completely confined within the single particle; this is the size-induced metal-insulator transition (SIMIT). The question arises as to whether the SIMIT for a finite system is continuous or discontinuous and at what critical particle size can one expect to see a SIMIT? The second situation relates to the electronic structure of ordered arrays of small particles, which are themselves individually metallic. By the controlled synthesis of such arrays in one, two or three dimensions, one can hope to engineer significant electron tunnelling and transport between neighbouring metallic particles so that the resulting electron wavefunction can ultimately be delocalized over macroscopic distances.

5.3.2 SIMITs in Isolated Clusters

5.3.2.1 Metal Clusters as Divided Metals

As shown in Figure 5.5, one way of analysing metal clusters, conceptually, is by starting from a macroscopic crystal of bulk metal and reducing the size of the crystal to the micron scale and ultimately to the nanometre scale. In this way, clusters can be regarded as small metallic particles or as very finely divided metals. As Perenboom has noted, starting with a metal crystal and gradually reducing its size, a metal to non-metal transition must occur, since an isolated atom cannot be described as a metal.

The phenomenon of discrete electronic energy level separation in small metallic particles was first highlighted by Fröhlich in 1937. However, the widespead interest in the study of small metallic particles stems from the work of Kubo, who proposed in the early 1960s that the discreteness of energy levels should lead to anomalies in the basic thermodynamic and electronic properties of small metallic particles at low temperatures.

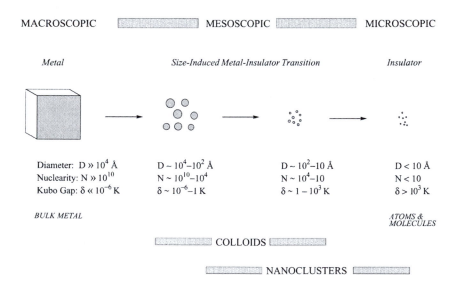

MACROSCOPIC   MESOSCOPIC   MICROSCOPIC

Metal   *Size-Induced Metal-Insulator Transition*   *Insulator*

Diameter: $D \gg 10^4$ Å	$D \sim 10^4$–10^2 Å	$D \sim 10^2$–10 Å	$D < 10$ Å
Nuclearity: $N \gg 10^{10}$	$N \sim 10^{10}$–10^4	$N \sim 10^4$–10	$N < 10$
Kubo Gap: $\delta \ll 10^{-6}$ K	$\delta \sim 10^{-6}$–1 K	$\delta \sim 1 - 10^3$ K	$\delta > 10^3$ K

BULK METAL   *ATOMS & MOLECULES*

COLLOIDS

NANOCLUSTERS

Figure 5.5 A representation of the successive division of a single grain of metal, leading eventually to a size-induced metal-insulator transition. The particle diameter (D), nuclearity (N) and Kubo gap (δ) are indicated, as are the size regimes spanned by metal clusters.

The Fröhlich–Kubo approach is based on the recognition that, in the limit of a bulk metal, electron energy levels are quasi-continuous and the electronic structure of the metal is well described by band theory. As the size of the metallic particle is reduced, however, the electronic energy levels become discrete – there are now a finite number of electronic states and the energy spacing between adjacent levels, for an *N*-atom particle is of the order E_F/N, where E_F is the Fermi energy of the bulk metal. The next step is to consider how the energy level spacing might be related to metallic or non-metallic behaviour (at non-zero temperature). DiCenzo and Wertheim have said that the existence of metallic properties depends on the existence of a partially filled band with small enough level spacing close to the Fermi level so that a small external potential (or thermal activation) can create mobile electron–hole pairs and allow a current to flow. This is a statement of the Kubo criterion or condition for a finite metallic particle (large cluster) to exhibit metallic conduction:

$$\delta \approx \frac{E_F}{N} \leq kT \tag{5.5}$$

where δ is the band gap or electronic energy level spacing at the Fermi level (defined as the highest occupied orbital in a finite particle). δ is also known as the Kubo gap. Smaller particles have larger δ values and hence require higher temperatures for metallic conduction. This transition from metallic to non-metallic behaviour, upon decreasing the size of a metallic particle, is shown schematically in Figure 5.6. The discretization of electronic energy levels, which leads to the suppression of metallic behaviour when $\delta > kT$, is responsible for the quantum size effects discussed in Chapter 1.

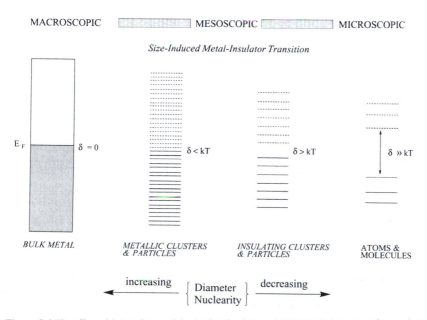

Figure 5.6 The effect of decreasing particle size (nuclearity) on the electronic structure of a metal. For finite metal particles, a Kubo energy gap opens up.

Table 5.1 The Kubo gap (δ) for sodium clusters as a function of diameter (D) and nuclearity (N).

Diameter (D/nm)	No. of atoms (N)	Kubo gap (δ/K)
10^3	1.6×10^{10}	5×10^{-6}
10^2	1.6×10^7	5×10^{-3}
10	1.6×10^4	5
1	16	5×10^3

Harrison and Edwards predicted the temperatures at which sodium clusters ($E_F(Na) = 3.24$ eV) of different sizes will be metallic according to Kubo's criterion. Thus, a cluster with a diameter D of 10 nm ($N \approx 16,000$) should behave as a metallic particle at $T > 5$ K, while a smaller cluster (e.g. $D \approx 5$ nm; $N \approx 2,000$) will be metallic at higher temperatures (50 K). For $N > 125$ atoms ($D > 2$ nm) sodium clusters should exhibit metallic conductivity at room temperature.

In Table 5.1, taking sodium clusters for example, we show the values for the diameter (D), number of atoms (N), and the Kubo gap (δ—in Kelvin) for sodium clusters with diameters in the range $1–10^3$ nm. The characteristic property of physical size (D, N) can thus be linked to the Kubo electronic energy gap, and the emergence of quantum size effects and the inevitable SIMIT. The macroscopic regime extends down to particle diameters of 1 micron ($N \approx 10^{10}$) and a negligibly small Kubo gap ($\delta \ll 10^{-6}$ K). The mesoscopic regime begins at the sub μm range, and terminates in the microscopic regime ($D < 1$ nm). Interestingly, the calculated Kubo gap for our prototypical metal, sodium, also provides an effective coarse-grained physical parameter for delineating the various electronic regimes of interest. Clearly, for bulk metals ($D \gg 1$ μm), we have $\delta \ll 10^{-6}$ K, and a continuum of electronic energy levels. Within the mesoscopic regime, we see the emergence of a rapidly increasing Kubo gap ($\delta \approx 10^{-6}–1$ K for $D \approx 10^3–10$ nm, to $\delta \approx 1–10^3$ K for $D \approx 10–1$ nm). Within this size regime one would clearly anticipate anomalies in the characteristic physical properties of particles and clusters arising from the onset of quantum-size effects. Equally, this is the same regime for which we might anticipate a SIMIT at relatively accessible temperatures (1 K and above). This is precisely the size regime of metal nanoclusters.

The Fröhlich–Kubo principle can be extended to transition metal particles, which are characterized by narrow d-bands and consequently high densities of electronic states. This leads to much narrower energy spacings between electron energy levels so that room temperature metallic conductivity should occur at smaller cluster sizes ($N > 40$) than for the s-valent alkali metals, such as sodium.

5.3.2.2 Measuring the Conductivity of an Isolated Cluster

Gor'kov and Eliashberg investigated the problem of the SIMIT in terms of the localization of the gas of conduction electrons in a metal through the (finite) size-induced confinement of the electron wave packet. They noted that, when the geometric dimensions of the metal particle or cluster become smaller than the characteristic phase coherence length of the conduction electrons, there should be a reduction in the (single-particle) d.c. electrical conductivity. Eventually, as the particle dimensions are continually reduced, there must be a SIMIT, due to the confinement (and collapse) of the conduction electron wave-packet—i.e. the electrons become localized.

Over thirty years ago, Ioffe and Regel predicted the occurrence of a metal-insulator transition1 in a macroscopic system when the criterion $k_F\ell_e = 1$ is satisfied (where k_F is the Fermi wave number and ℓ_e is the mean free path of the conduction electrons). Wood and Ashcroft later noted that if the electron mean free path is regarded as the characteristic dimension (i.e. the diameter, D) of a small particle or cluster, then the Ioffe–Regel criterion may be modified to give:

$$k_F\ell_e = k_F D \approx 1 \tag{5.6}$$

in order to describe the cessation of metallic (conducting) behaviour within an individual particle.

Measuring the d.c. conductivity of isolated metal particles, of course, is arguably the most direct method of investigating the size-induced metal-insulator transition within a single particle. Marquardt and Nimtz have succeeded in measuring the quasi-d.c. conductivity of individual indium nanoparticles by performing contactless microwave absorption experiments for metal particles with diameters ranging from 10 nm to of the order of microns. They found that the electrical conductivity within each particle (σ) decreases rapidly with decreasing particle diameter, below a few μm, according to the approximate relationship $\sigma \propto D^3$, corresponding to a SIMIT.

Recent microwave loss measurements by Edwards and co-workers, made on colloidal suspensions of gold particles in a microwave cavity, has shown that gold clusters with $D \approx 3$ nm essentially exhibit no conductivity ($\sigma \approx 10^{-6}$ $\Omega^{-1}m^{-1}$). The conductivity rises abruptly for $D \gtrsim 10$ nm, corresponding to an insulator-metal transition at around 10,000 atoms. Clusters with diameters of approximately 50 nm show relatively high conductivities ($\sigma \approx 3\times10^7$ $\Omega^{-1}m^{-1}$), as evidenced by high, broad band microwave loss. In fact colloidal suspensions of gold particles of around this size form the basis for some of the paints used by the military for 'stealth' technology, whereby aeroplanes are effectively made invisible to radar detection. Clusters in the micron size range have higher conductivities, approaching that of bulk gold ($\sigma \approx 6\times10^7$ $\Omega^{-1}m^{-1}$). However, since their diameters are significantly greater than the skin depth of the microwave radiation, their microwave absorption is relatively low and their main effect is to reduce the effective size of the microwave cavity, leading to an increase in the cavity resonance frequency.

Scanning tunnelling spectroscopy (STS) represents an extremely powerful technique for the direct measurement of the *I–V* (current vs. voltage) characteristics of individual clusters and particles deposited on a surface (usually a metal or graphite), and hence for investigating the metallic or insulating nature of individual particles. Schmid and colleagues first used STS to study the *I–V* behaviour of passivated Pt_{309} clusters on a gold substrate (see Chapter 8). Rao and co-workers have demonstrated that bare gold clusters with diameters larger than 4 nm show finite currents for very small applied voltages, implying genuine metallic character within the particle (with a resistance which obeys Ohm's law). In contrast, clusters with diameters of around 1.4 nm exhibit rectifying behaviour, which is consistent with insulating status within the particle and a so-called Coulomb gap. STS measurements on particles of Pd, Ag, Cd and Au by the Rao group have shown that there is a general marked decrease in the measured conductance for diameters below 1 nm, at which sizes the measured density-of-states exhibits a band gap.

Rao *et al.* have examined the cluster size-dependence of the slope of the *I–V* curves in some detail. As the cluster size increases, the normalized conductance increases linearly up to a volume of some 4 nm^3 ($D \approx 2$ nm) and reaches a constant value for larger particles and clusters. The measured value of the conduction gap lies between 10 mV and 70 mV and decreases with increasing cluster volume until there is no conduction gap for clusters with diameters greater than 1 nm. This observation suggests that small clusters, with diameters below 1 nm, are indeed insulating, in that the extrapolated zero-temperature d.c. conductivity is zero. These results are consistent with calculations by Rosenblit and Jortner who predicted electron localization to occur within a single metal cluster for diameters $D \leq 0.6$ nm.

Relativistic DFT calculations on gold clusters, by Rösch, have indicated a decrease in HOMO–LUMO gap from 1.8 eV for Au$_6$ ($D \approx 0.5$ nm) to 0.3 eV for Au$_{147}$ ($D \approx 2$ nm), though other *ab initio* calculations have predicted a smaller gap of 0.1 eV for this larger cluster. These results are qualitatively consistent with the microwave and STS measurements described above, which indicate the onset of metallic behaviour (at room temperature) for gold particles in the nanometre size regime. The consequences of size-dependent conductivity behaviour for possible single-electron tunnelling devices, is discussed further in Section 8.2.

5.3.2.3 The Non-metal to Metal Transition in Mercury Clusters

A series of important experiments probing the electronic structure of individual (gaseous) metal clusters and particles was performed by Rademann, Hensel and co-workers in the early 1980s. These authors interpreted the size-dependent variation of the ionization energies (IP) of mercury clusters (reproduced in Figure 5.7) in terms of a continuous transition from van der Waals to metallic bonding in the region of $N = 13$–70 atoms. For $N < 13$, the approximate straight line behaviour of ionization energy (IP) vs. $1/R$ (where R is the cluster radius, assuming a spherical cluster geometry—see Section 4.2.1) extrapolates to a bulk value of IP(∞) ≈ 6.5 eV. This extrapolated value is significantly higher than the bulk work function of elemental Hg (4.49 eV) and reflects a fundamentally different type of bonding in these small clusters. For $N > 13$, the measured IP decreases more rapidly and converges on the straight line predicted by the Liquid Drop Model at $N \approx 140$.

The closed sub-shell electronic configuration, $6s^2$, of the free Hg atom causes small mercury clusters to be non-metallic and held together by relatively weak van der Waals dispersion forces (as found, for example, in noble gas clusters). As the cluster grows, the atomic 6s and 6p levels broaden into electronic energy bands (see Figure 5.8). A metal-insulator transition within the cluster is presumed to occur at a critical nuclearity (N_c) due to 6s–6p band overlap (as shown in Figure 5.8), though there is probably a transition to a semiconducting (covalently bonded) state (due to s-p band hybridization, prior to band overlap) before the metallic state is reached.

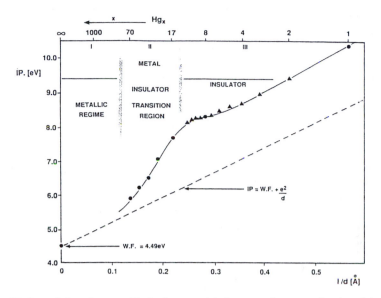

Figure 5.7 The variation of measured ionization potential of mercury clusters as a function of cluster size. The work function for bulk Hg (4.49 eV) is indicated. The dashed line is the calculated IP—based on the classical LDM.

For an isolated cluster, the individual s- and p-band widths (W_s and W_p respectively) are related to the mean atomic coordination number (Z_{av}) by:

$$\left(W_s, W_p\right) \propto Z_{av}^{1/2} \tag{5.7}$$

where the square root dependence derives from a tight binding analysis. The metal–insulator transition within the cluster can therefore be viewed as occurring at a critical mean coordination number (Z_{av}) and, implicitly a critical nuclearity (N_c), rather than a critical density (ρ_c), which is the situation in the macroscopic condensed phase system.

Building upon an analogy with expanded liquid mercury, where a density-induced metal-insulator transition occurs at $\rho_c = 5.75$ gcm^{-3}, corresponding to $Z_{av} \approx 6$–7, Tománek and co-workers predicted a transition to the metallic state for Hg$_N$ clusters in the range $20 \le N \le 50$. Tight binding calculations by Pastor and Bennemann predicted a fundamental change in chemical bonding from van der Waals to covalent at around $N = 13$, with a transition to metallic bonding at around $N = 80$, in good agreement with the ionization energy measurements of Rademann and Hensel.

Pastor and Bennemann also considered the detailed process of electronic delocalization within an isolated cluster as the nuclearity increases. When the electrons delocalize within a cluster, they jump from one atom to one of its neighbours and pairs of ions (Hg$^+$:Hg$^-$) form instantaneously. This intracluster electron hopping process is enhanced by the $6s$ electronic bandwidth arising from the interaction of the Hg atoms (see Figure 5.8).

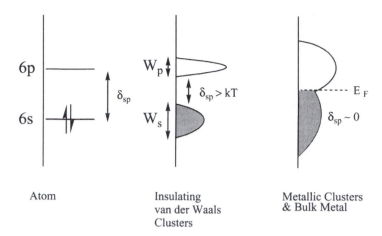

Figure 5.8 The transition from atomic to metallic electronic structure with increasing cluster size, for mercury clusters. Occupied bands are indicated by shading.

Pastor and Bennemann have proposed an important criterion for electron delocalization within the cluster, which corresponds closely to Mott's criterion for the metal–insulator transition in macroscopic systems, reflecting competition between kinetic and Coulomb (charging) energy terms. As in Mott's theory, delocalization occurs when a critical electronic bandwidth and efficient screening are achieved. In the case of clusters, however, the critical electronic bandwidth is related to a critical coordination number, instead of a critical interatomic distance (i.e. critical density).

There are a number of other experiments on mercury clusters, such as studies of the size-dependence of the $5d$–$6p$ autoionization spectrum and the appearance of the surface plasmon mode, which provide supporting evidence for an insulator to metal transition beginning at around 13–20 atoms. Photoelectron spectroscopic measurements for mass selected Hg_N^- clusters ($N = 3$–250) show that HOMO–LUMO gaps (for the neutral clusters) decrease gradually from 3.5 eV for Hg_3 to 0.2 eV for Hg_{250}, with band-gap closure predicted to occur at $N \approx 400$. The seeming disagreement between these values and the results of the IP measurements may be due to differences in electronic energy levels (due to differing electron–electron repulsion) between the species Hg_N, Hg_N^+ and Hg_N^-.

5.3.3 Metallization of Arrays of Particles

Recent developments in the synthesis of small metallic particles have involved the production of relatively monodisperse, organically-functionalized colloidal metal particles, and their crystallization to form ordered arrays (see Section 8.2). In principle, chemical control over the composite particle size, and the inter-particle separation provides a direct route for investigating the coupling between individual particles. Indeed two dimensional, ordered arrays of such assemblies have been prepared at the air/water interface on a Langmuir trough, and the Langmuir technique can be used to tune the interparticle separation continuously.

Heath, Saykally and co-workers have recently reported the observation of a reversible, room-temperature metal–non-metal transition in organically-functionalized silver particle monolayers. A useful parameter for characterizing these monolayers is the quantity (d/D), where d is the interparticle separation, as measured between particle centres, and D is the particle diameter (as shown schematically in Figure 5.9).

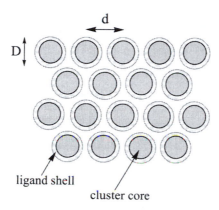

Figure 5.9 Schematic representation of a hexagonal film of thiol-passivated metal clusters, showing the two important length scales—the cluster diameter (D) and the inter-cluster (centre–centre) separation (d).

The UV–visible reflectance spectra from a Langmuir–Blodgett (LB) film of 4 nm diameter silver particles, collected *in-situ* as the film is compressed, is shown in Figure 5.10. Upon initial compression, the film becomes more reflective (see (1) (2) and (3) in Figure 5.10). The significant decrease in reflectivity, which occurs between (2) and (3) corresponds to a disconuity, due to the onset of Second Harmonic Generation (SHG)—a non-linear optical effect. The final reflectance spectrum is similar to that reported for thin, metallic silver films ((4) and (5) in Figure 5.10) and indicates conclusively that the LB film has become metallic. The compressibility of the film arises due to interdigitation of the organic ligands and to deformation of the ligands themselves.

A reasonable working picture highlights the energy gain from electron delocalization; when this energy gain becomes sufficient to overcome the site (single-particle) charging energy, there is an insulator → metal transition throughout the entire array of particles. The experimental data provide strong evidence for the onset of strong quantum mechanical interactions between adjacent silver particles (via tunnelling through the intervening ligands) when the interparticle separation distances are reduced below 1.2 nm. The reversible insulator → metal transition occurs when the interparticle separation is reduced below 0.5 nm. These studies have led to the interesting proposal that this reversible metal–insulator transition represents a first-order Mott transition at room temperature.

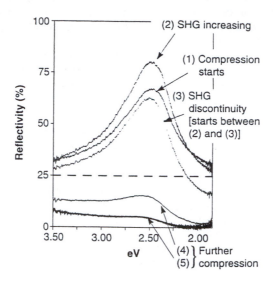

Figure 5.10 UV–visible reflectance spectra of a LB monolayer of 40 Å diameter silver clusters, as the film is compressed (1-5).

Earlier studies by Kreibig and co-workers also impact on the nature of the insulator-metal transition in extended arrays of mesoscopic particles. These authors investigated the effect of particle aggregation by monitoring the extinction spectrum from gold clusters all the way up to a thin film of gold. The degree of aggregation has a drastic effect on the optical response; not only are the spectral intensities changed but the qualitative nature of the spectrum is altered. On increasing aggregation, the spectrum is continuously altered from the relatively narrow plasma absorption of a mesoscopic particle to the broad extinction features which resemble the optical absorption of a continuous thin film of gold.

5.4 CLUSTER MAGNETISM

5.4.1 Magnetism

The conduction electrons in a metal are responsible for the characteristic finite (but small) magnetic susceptibility of simple bulk metals—so-called Pauli paramagnetism—since in the presence of an external magnetic field they acquire a net inbalance of up- and down-spins. Since $E_F \gg kT$ for most accessible temperatures and the number of spins contributing to the magnetism is very small, the Pauli paramagnetism for macroscopic bulk metals is usually weak and temperature-independent. For particles with diameters less than 10 nm, Kubo predicted that, due to the emerging discreteness of the electronic energy levels of particles of finite size, one should now be able to distinguish between particles having odd or even numbers of conduction electrons. The odd-number small particles of a monovalent atom (e.g. Na) would thus exhibit Curie paramagnetism in the magnetic susceptibility at low temperatures, rather than the Pauli paramagnetism characteristic of the bulk metal. The resulting spin-pairing of electrons in even-number particles at low temperatures also tends to reduce their intrinsic magnetic susceptibility below the characteristic Pauli value. These anticipated changes in magnetic behaviour are just one example of quantum size effects arising from situations in which the Kubo gap (δ) is comparable to, or greater than, the thermal energy (kT) at any particular temperature.

The d-electrons of a transition metal atom are fairly well localized and, as transition metal atoms may have up to five unpaired electrons with parallel spins, large magnetic moments are possible. Because of this, many transition metal elements exhibit long range ordering of the atomic magnetic moments. For metals and alloys, this long range magnetic ordering usually conforms to one of two types: *ferromagnetism*—where magnetic moments on neighbouring atoms are aligned in a parallel fashion (as in bulk iron, cobalt and nickel); and *antiferromagnetism*—where the neighbouring magnetic moments are arranged in an antiparallel fashion (as in bulk chromium).

5.4.2 Magnetism of Carbonyl Clusters

Johnson and co-workers carried out a systematic investigation of magnetism in osmium carbonyl clusters $[Os_N(CO)_x]^{z-}$. The experimental quantity of most direct interest was the *excess molecular susceptibility* (χ_{ems}) of the Os cluster compounds, which represents the difference between the experimentally-determined magnetic susceptibility of the cluster and the sum of the diamagnetic corrections of those species from which the high-nuclearity cluster is composed. These excess molecular susceptibilities are most naturally interpreted as arising from a Van Vleck (temperature-independent) paramagnetic contribution to the magnetic susceptibility. The temperature-independent Van Vleck paramagnetism was found to increase sequentially with cluster size as the molecular electronic states begin to approach the electronic band or continuum regime, which ultimately results in the characteristic Pauli paramagnetism in the bulk, metallic state.

Johnson and Edwards also addressed the question of how many atoms constitute a metal—finding that the answer depends critically upon the particular property being measured. Transport properties (e.g. d.c. electrical conductivity) require, as a prerequisite for the detection of metallic character, the complete ionization of an electron from its parent site, which typically involves energy changes in the range 1–10 eV. In contrast, as far as magnetic properties are concerned, modifications to the intra- and inter-unit exchange energy in the developing metal cluster need only be of the order of magnitude of the magnetic Zeeman or hyperfine energies (both typically in the range 10^{-3}–10^{-4} eV) for changes to be detectable.

5.4.3 Superparamagnetism in Free Transition Metal Clusters

The magnetism of free clusters of ferromagnetic $3d$ transition metal elements, (iron, cobalt and nickel) have been studied by measuring the deflection of a molecular beam of clusters in an inhomogeneous magnetic field. This experiment is a modification of that devised by Stern and Gerlach over 70 years ago to measure the spin of the electron. In the cluster magnetism measurements of de Heer and colleagues, clusters are generated by pulsed laser vapourization of a metal target, the metal vapour is rapidly quenched by a high-pressure (10^6 Pa) jet of helium, before expansion into a low-pressure (10^{-5} Pa) region, creating a supersonic cluster beam, which then passes through a gradient magnet before detection by a TOF mass spectrometer.

In the original Stern–Gerlach experiment, a beam of silver atoms was split into two equal components, corresponding to atoms with their electronic spin angular momenta pointing 'up' or 'down' (i.e. with quantum numbers $m_s = +\frac{1}{2}$ or $-\frac{1}{2}$). Independent experiments by de Heer and Bloomfield showed, however, that magnetic transition metal clusters are only deflected in one direction, as shown in Figure 5.11. This difference arises because clusters have more degrees of freedom available than do atoms. Coupling of spin angular momentum with the rotational angular momentum of the cluster as a whole (and with cluster vibrational modes) serves to align the average magnetic moment of the cluster with the field, leading to deflection in one direction only. This also implies that the magnetic moments are dynamically decoupled from the lattice though they are dynamically coupled together via exchange interactions. Thus, the magnetic moments rotate in unison relative to the lattice, under the influence of the applied field, and the cluster behaves as a paramagnet with a large net magnetic moment—i.e. a 'superparamagnet', as shown schematically in Figure 5.12a. At temperatures above a certain, critical temperature (T_c—the counterpart of the Curie temperature in bulk ferromagnets), thermal fluctuations are sufficiently large to overcome the coupling of spins with the applied field and there is no superparamagnetic moment (Figure 5.12b).

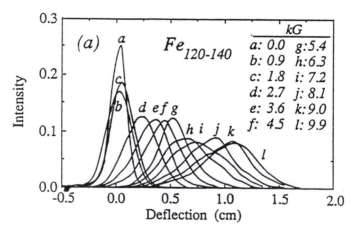

Figure 5.11 Stern-Gerlach profiles of Fe$_N$ clusters ($N = 120$–140) for various applied magnetic fields.

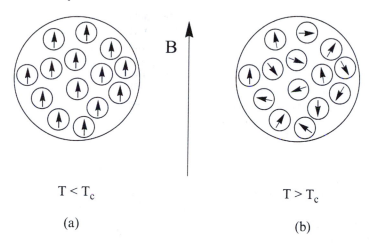

$$T < T_c \qquad\qquad\qquad T > T_c$$

$$\text{(a)} \qquad\qquad\qquad \text{(b)}$$

Figure 5.12 (a) Superparamagnetic alignment of electron spins in an Fe, Co or Ni cluster in the presence of a magnetic field (B), at $T < T_c$. (b) Thermal randomization of spins for $T > T_c$.

The effective magnetic moment (μ_{eff}, and hence the magnetization $M = N\mu_{\text{eff}}$) is a time average of the projection of the superparamagnetic moment ($\mu_{\text{super}} = N\mu$, where μ is the internal magnetic moment per atom):

$$\mu_{\text{eff}} = \mu L\left(\frac{N\mu B}{kT}\right) \tag{5.8}$$

where B is the applied magnetic field, T is the vibrational temperature of the cluster and L is the Langevin function ($L(x) = \coth(x) - 1/x$). Although the magnetization per atom of $3d$ transition metal clusters is found to be smaller for clusters than for the corresponding bulk, the magnitude of the superparamagnetic moment (and hence the average magnetic moment per atom) is actually found to be higher than that of the bulk (e.g. $\mu(\text{Fe}_\infty) = 2.2$ μ_B, $\mu(\text{Co}_\infty) = 1.7$ μ_B, $\mu(\text{Ni}_\infty) = 0.6$ μ_B). In smaller clusters the d band is narrower and there is therefore less d electron delocalization. Since this delocalization is responsible for the reduction of μ from the free atom value ($\mu(M_1) = 3$, 2 and 1 μ_B for $M =$ Fe, Co and Ni, respectively), the cluster has an internal magnetic moment per atom which is intermediate between the free atom and bulk values: $\mu(M_1) > \mu(M_N) > \mu(M_\infty)$.

Experiments have shown that the average magnetic moment per atom decreases as the cluster gets larger, reaching the bulk limit at around 500 atoms for Fe. Oscillations in magnetic moments have been attributed to geometric and electronic shell effects (see Chapter 4), as well as charge- and spin-density waves.

The $4d$ transition metal rhodium (Rh), which is not magnetic in the solid state, has actually been shown (by Bloomfield and co-workers) to form superparamagnetic clusters! This is an interesting example of a finite size effect, where a cluster has a property which is distinct from that of the bulk. Interestingly, ruthenium and palladium clusters (Ru and Pd are either side of Rh in the periodic table) do not exhibit superparamagnetism. Knickelbein has shown that manganese clusters Mn_N ($N = 11$–99) are also superparamagnetic (like those of Fe, Co and Ni), despite the fact that no ferromagnetic phase of bulk manganese is known. This finite size effect may be caused by the clusters having different structures and bond lengths (and hence different Mn–Mn coupling constants) compared to the bulk. Local minima in magnetic moment (per atom) for Mn_{13} and Mn_{19} have been interpreted in terms of an icosahedral growth mode. For larger clusters, broad oscillations are observed in the magnetic moments per atom—with a minimum around Mn_{32-37} and a maximum around Mn_{50-56}. These variations may be due to changes in structure type at these nuclearities.

Bloomfield and co-workers have also looked at the magnetism of clusters of the rare earth (lanthanide) elements, where the magnetism is due to strongly localized $4f$ electrons. The magnetic behaviour of these clusters is more complicated than that of the transition metals—some exhibit superparamagnetism, while others display 'locked moments' because the coupling of the magnetic moments to the lattice is greater than kT, and thus they are stable to thermal fluctuations and behave as permanent magnets (e.g. ferromagnets). Such clusters have smaller internal magnetic moments (per atom) than in the bulk, with the magnetic moment rising with increasing cluster size.

Finally, Knickelbein has shown that adsorption of a single CO molecule on Ni_N clusters ($N = 8$–18) leads to a significant reduction (quenching) of the magnetism of certain clusters, such as Ni_8, Ni_9, Ni_{15} and Ni_{18}—for example the total magnetic moment of Ni_8 is reduced by 5 μ_B—while others (e.g. Ni_{12}, Ni_{16} and Ni_{17}) are almost unchanged. Magnetic quenching is predicted by simple models based on electron localization accompanying cluster–ligand bond formation—but such simple models do not account for the large quenching observed for Ni_8 or the lack of quenching for Ni_{12} and further, more detailed theoretical studies are necessary to explain these experimental observations.

5.4.4 Magnetism in Simple Metal Clusters

Stern–Gerlach deflection experiments (similar to those described above for transition metal clusters) have been performed by Cox and co-workers on molecular beams of neutral aluminium clusters. They found that deflections decrease with increasing cluster size, being unmeasurably small for $N > 12$. The odd-electron (i.e. odd-atom) clusters have only one unpaired electron (i.e. a doublet ground state), while even-electron clusters either have no or two unpaired electrons, giving rise to singlet and triplet ground states, respectively. Triplet ground states were identified for Al_2, Al_6 and Al_8.

An alternative method for investigating the number and distribution of unpaired electrons in a cluster is *electron spin resonance* (ESR—also known as *electron paramagnetic resonance*) spectroscopy. In such experiments, clusters are deposited on an inert matrix, which is then placed within a microwave cavity in a homogeneous magnetic field. The field is varied until resonance is reached, when the energy gap between the $m_s = -\frac{1}{2}$ and $m_s = +\frac{1}{2}$ spin states of the unpaired electron matches the energy of the microwaves, then a spin-flip transition occurs. Information about the structure of the cluster is obtained by analysis of the hyperfine structure—due to coupling between the magnetic moment of the unpaired electron and that of the nuclei (if they have non-zero nuclear spin) of the atoms in the vicinity of the electron.

Molecular orbital calculations on the Li_7 cluster predict a pentagonal bipyramidal geometry, with a single unpaired electron lying in a non-degenerate axially symmetric a_2'' orbital. The ESR spectrum of isotopically pure $^7Li_7$ clusters has been measured by Garland and Lindsay, with the clusters being deposited in an inert matrix. The spectrum consists of a single very complicated multiplet, but it has successfully been analysed and found to be consistent with the MO calculations. The ESR signal due to the unpaired electron in the a_2'' orbital is split by coupling to the two axial atoms (as each ^7Li nucleus has a nuclear spin $I = 3/2$, the maximum total nuclear spin of the two axial atoms $I_{max} = 3$) to give seven ($2I_{max} + 1$) lines, with a coupling constant $J_{ax} = 40$ G. Each of these lines is then split by coupling with the five equatorial atoms ($I_{max} = 15/2$) into sixteen lines, with a coupling constant $J_{eq} = 2$ G. The difference between the axial and equatorial coupling is also consistent with the MO calculations, as the a_2'' orbital is calculated to be mostly localized on the axial Li atoms.

5.4.5 Giant Magneto-Resistance in Granular Metals

There has been much recent interest in granular materials formed by embedding clusters in a solid host. When embedded in metals, or even insulators, magnetic clusters (e.g. of Cr, Fe, Co, Ni or mixtures of these metals) are known to exhibit the phenomenon of *Giant Magneto-Resistance* (GMR), with a change in resistance upon application of a magnetic field as high as 20% for Co clusters embedded in Ag. Such magneto-resistive materials show considerable promise for applications as magnetic sensors and for magnetic recording.

Another interesting area of research is the study of one-, two- and three-dimensional arrays of surface-passivated (see Section 8.2) superparamagnetic clusters, which may have applications in magnetic memory storage. In view of Knickelbein's results, however, it is possible that the process of chemical passivation will be accompanied by a loss of superparamagnetism.

5.5 EXERCISES

5.1 Discuss the ways in which cluster reactivity experiments can give information on cluster structure.

5.2 Use Equation (5.5) to predict the temperature at which a 1,000-atom sodium cluster should become metallic. (The Fermi energy of sodium, $E_F(Na) = 3.24$ eV.)

5.3 What is the evidence for the change of mercury clusters from van der Waals to covalent (semiconductor) to metallic clusters with increasing size?

5.4 What is the origin of superparamagnetism in iron and nickel clusters?

5.6 FURTHER READING

Bönnemann, H. and Brijoux, W., 1999, Potential applications of nanostructured metal colloids. In *Metal Clusters in Chemistry*, Vol.2, edited by Braunstein, P., Oro, L.A. and Raithby, P.R., (Weinheim, Wiley-VCH), pp. 913-931.

Edwards, P.P., Johnston, R.L. and Rao, C.N.R., 1999, On the size-induced metal-insulator transition in clusters and small particles. In *Metal Clusters in Chemistry*, Vol.3, edited by Braunstein, P., Oro, L.A. and Raithby, P.R., (Weinheim, Wiley-VCH), pp. 1454–1481.

Jellinek, J., 1996, Theoretical dynamical studies of metal clusters and cluster-ligand systems. In *Meto'   l Interactions*, edited by Russo, N. and Salahub, D.R., (Dordrecht, Klu   p. 325–360.

Johnston, R.L., 19   e development of metallic bonding in clusters. *Philosophical T   tions of the Royal Society of London*, A356, pp. 211–230.

Jortner, J., 19°   ter size effects. *Zeitschrift für Physik*, D24, pp. 247–275.

Knickelbei   ., 1999, Reactions of transition metal clusters with small molec   nnual Reviews of Physical Chemistry*, 50, pp. 79–115.

Krei   and Quinten, M., 1994, Electromagnetic excitations of large clusters. *   sters of Atoms and Molecules*, Vol.II, edited by Haberland, H., (Berlin, inger-Verlag), pp. 321–359.

astor, G. M. and Bennemann, K. H., 1994, Transition from van der Waals to metallic bonding in mercury clusters. In *Clusters of Atoms and Molecules*, Vol.I, edited by Haberland, H., (Berlin, Springer-Verlag), pp. 86–113.

Rao, C.N.R., Kulkarni, G.U., Thomas, P.J. and Edwards, P.P., 2000, Metal nanoparticles and their assemblies. *Chemical Society Reviews*, 29, pp. 27–35.

Riley, S.J., 1994, Chemistry with neutral metal clusters. In *Clusters of Atoms and Molecules*, Vol.II, edited by Haberland, H., (Berlin, Springer-Verlag), pp. 221–240.

Schmid, G., 1999, Nanosized clusters on and in supports—perspectives for futurecatalysis. In *Metal Clusters in Chemistry*, Vol. 3, edited by Braunstein, P., Oro, L.A. and Raithby, P.R., (Weinheim, Wiley-VCH), pp. 1325–1341.

Shi, J., Gider, S., Babcock, K., Awschalom, D.D., 1996, Magnetic clusters in molecular beams, metals and semiconductors. *Science*, 271, pp. 937–941.

CHAPTER SIX

Semiconductor Clusters

6.1 INTRODUCTION

In this chapter, the term 'semiconductor cluster' will be used to denote a cluster of an element or compound, which in the bulk, under ambient conditions, is a semiconductor—i.e. a solid with a covalently bonded network of atoms and a small gap between the occupied (valence) and empty (conduction) bands of levels, such that electrons can be promoted thermally or upon absorption of light, leading to conduction. The semiconducting elements are mostly found in groups 13–16 of the periodic table. Semiconductor compounds are formed by taking combinations of these elements (though 12–16 or II–VI semiconductors, such as cadmium selenide, are also of technological importance). In this chapter, I will concentrate on elemental semiconductor clusters of group 14 (C, Si, Ge and Sn) and the isoelectronic 13–15 compounds (e.g. GaAs and InP). The more polar metal chalcogenide clusters (e.g. ZnS and CdSe) will be discussed in Chapter 7.

6.2 CARBON CLUSTERS

For the lighter group 14 elements (carbon, silicon and germanium), polyatomic species have been identified in the equilibrium vapour above the heated solid. This is in contrast to simple metals (see Chapter 4), where the vapour is primarily monoatomic. This tendency to form clusters in the equilibrium vapour is due to the strong, directional covalent bonding exhibited by these elements. Phillips has described these clusters as 'network clusters', by analogy with their solids, which can be considered as infinite networks of covalent bonds.

Carbon forms clusters more readily than any other element. These clusters possess strong covalent bonds and many have both σ- and π-bonding components. Carbon is one of the most abundant elements in the Universe: carbon-rich red giant stars emit enormous amounts of carbon dust into the interstellar medium—with carbon grains constituting the major component of interstellar dust. The largest molecules that have so far been identified in space (on the basis of their microwave spectra) are long carbon chains such as $HC_{11}N$. Probably the most famous clusters of all are carbon clusters—the fullerenes, which have attracted so much attention and engendered so much excitement since their discovery in the late 1980s.

6.2.1 Small Carbon Clusters

The first observation of carbon cluster cations, C_N^+ (with $N \leq 15$) was made by Hahn and co-workers as long ago as 1940. The clusters were produced in a high frequency discharge between graphite electrodes and detected in a mass spectrometer. A more detailed investigation was made by Hintenberger in the late 1950s. In these experiments, an odd-even periodicity (odd N having higher intensity than even N) was observed in the mass spectral intensity up to $N = 10$, after which an intensity fluctuation with a periodicity of four carbon atoms was observed. Similar results were subsequently obtained (in the 1980s) in for carbon cluster cations generated by laser vaporization of a carbon foil. The intensity variation is more marked if the laser-vaporized carbon is cooled by a pulse of a rare gas, such as helium—with magic numbers for C_N^+ at $N = 7$, 11, 15, 19 and 23 being particularly marked. The fine details of the mass spectra, in particular, below $N = 10$, are sensitive to experimental conditions.

Neutral cluster distributions have been investigated by UV photoionization of the neutral clusters generated by laser vaporization. The mass spectral intensities are very similar to those for the cluster cations which are generated in the vaporization process itself. This similarity is probably because small carbon clusters all have ionization energies greater than 7 eV, meaning that ionization (using for example an ArF excimer laser with a photon energy of 6.4 eV) is a multi-photon process, which is accompanied by fragmentation. Detailed photofragmentation studies of C_N^+ clusters have revealed that the primary fragmentation channel involves loss of neutral C_3, though larger clusters also exhibit loss of neutral C_5.

In 1959, on the basis of molecular orbital calculations, Pitzer and Clementi predicted that the neutral carbon species C_N should be stable as chains for $N \leq 9$, but that there should be a transition to ring structures for higher nuclearities. Recent, more sophisticated calculations have confirmed a chain-ring crossover at around $N = 10$. These calculations have shown that neutral linear clusters with an odd number of C atoms are energetically favoured up to $N = 10$, due to orbital filling considerations. The linear structures of C_3 and C_5 have been confirmed by high resolution spectroscopic measurements and these laboratory spectra have been used to identify astronomically detected spectral features due to C_3 and C_5. The spectral lines due to the C_3 molecule were first observed in 1881 in the spectrum from a comet, though they were not identified until 1951. More recently, C_5 has been detected in the infrared spectrum of a carbon star, by Bernath and co-workers.

Recent calculations on small even-membered carbon clusters have indicated that for some (such as C_4) the ring structure may actually be lower in energy than the chain at very low temperatures. However, at the temperature of cluster generation the chain isomer is favoured due to its higher vibrational entropy (since chains are floppier than rings, they have more low-frequency vibrations). This same argument can be invoked to rationalize the appearance of chains for higher nuclearity clusters under experimental conditions where the clusters are generated with very high internal energies.

Calculations on ring clusters predict that even-membered double-bonded rings, with $N = 4k+2$ atoms, should be favoured. (This $4k+2$ rule is analogous to the Hückel $4n+2$ rule for aromatic cyclic hydrocarbons.) The experimental observation of high mass spectral intensities for odd-carbon species, $N = 11, 15, 19, 23$, etc., can be explained by considering the ionization energies, which are lower for the open shell odd-carbon clusters, and the fact that photofragmentation of a C_N^+ ring with $N = 4(k+1)+2$ atoms will, after loss of the stable C_3 molecule, lead to a fragment with $N-3 = 4k+3$ (i.e. 11, 15, 19,...) atoms.

Bowers and co-workers have used drift tube experiments to measure the mobilities of carbon cluster ions and, hence, to deduce their topologies (see Section 1.3.5.2). In this way, they managed to observe the transition from linear chains to monocyclic and bicyclic rings and eventually to fullerenes as a function of cluster size.

6.2.2 The Fullerenes

6.2.2.1 The Discovery of C_{60}

In the early 1980s, Kaldor and co-workers at EXXON used a laser vaporization source to generate carbon clusters and detected cluster cations with up to 100 atoms by mass spectroscopy. Similar experiments were carried out, at around the same time, by a group at AT&T Bell laboratories. The mass spectrum obtained by Kaldor *et al.*, which is reproduced in Figure 6.1, shows two distinct sets of features.

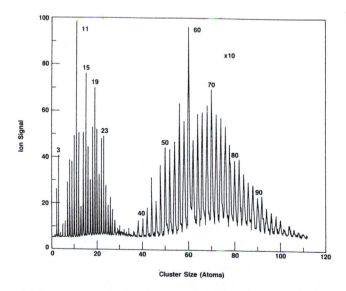

Figure 6.1 Mass spectrum of carbon clusters produced from a laser vapourization source.

The first set, corresponding to small clusters ($N \leq 30$), has peaks for both even- and odd-atom clusters and has been discussed in the previous section. There is then a gap or region with few significant peaks until a second set of peaks is observed for $N \geq 38$. In this region of larger clusters, it is noticeable that there are only peaks for clusters with even numbers of atoms ($N = 2k$). This even-only pattern (which in fact starts at $N = 30$) was wrongly attributed to a carbyne phase, with clusters composed of cross-linked acetylenic ($-C\equiv C-)_n$ chains. No particular relevance was assigned to the relatively intense peak due to C_{60}^{+}, which can be seen in Figure 6.1.

In 1985 Harry Kroto visited the laboratory of Rick Smalley at Rice University in Texas, with the aim of simulating carbon star chemistry using Smalley's laser vaporization–helium carrier gas cluster source. Although they were initially interested in the smaller long-chain cluster species, Kroto, Smalley, Curl and co-workers became intrigued by the C_{60} peak, the intensity of which they found varied greatly depending on the source conditions. Indeed, under conditions favouring increased clustering, the C_{60} peak was found to dominate the mass spectrum—the only other significant peak being due to a C_{70} species—as shown in Figure 6.2.

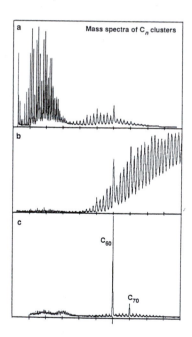

Figure 6.2 Mass spectra of carbon clusters produced by laser vaporization, showing the effect of increasing clustering (**a–c**).

After much discussion, Kroto, Smalley and co-workers came to the conclusion that a likely structure for the apparently very stable C_{60} cluster should have high symmetry and the structure should resemble that of graphite, which consists of a hexagonal network of carbon atoms. The now-famous structure that they came up with, which is shown in Figure 6.3a, is a hollow truncated icosahedron, which does indeed have very high symmetry (I_h)—so high that all 60 atoms are equivalent by symmetry in this structure! The cluster polyhedron is composed of twelve pentagonal rings (or faces) and twenty hexagonal rings/faces. The structure seemed familiar to its creators and, indeed, it turns out that the pentagons and hexagons define the pattern commonly found on a football (or 'soccer ball' in the USA). The Kroto–Smalley–Curl group christened the C_{60} cluster 'buckminsterfullerene' by analogy with the beautiful, highly symmetric geodesic domes designed by the American architect Robert Buckminster Fuller. The name 'fullerenes' has subsequently been given to the homologous family of carbon clusters.

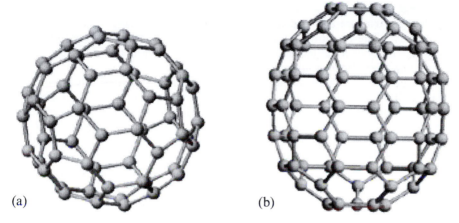

(a)   (b)

Figure 6.3 Structures of the two most abundant fullerenes produced in laser vaporization experiments: (a) C_{60} (I_h); (b) C_{70} (D_{5h}).

In fact, Kroto, Smalley and Curl were unaware that, as long ago as 1966, David Jones (writing under his pseudonym of Daedalus in *New Scientist*) had postulated hollow-shell graphite molecules and had even predicted (sometimes correctly) their properties. In 1970, Ōsawa actually proposed the truncated icosahedral C_{60} cluster as a possible super aromatic molecule! Independently, Bochvar and G'alpern (in 1973) performed Hückel calculations on truncated icosahedral C_{60}, obtaining a correct description of the π-electron system (see Section 6.2.2.4).

The beautifully simple structure proposed for C_{60}, and its relationship to the graphite structure, led many researchers to study carbon clusters. Although the chemists Kroto, Smalley and Curl spent five years trying to generate sufficient quantities of C_{60} to confirm its structure, they were eventually pipped at the post by two physicists, Krätschmer and Huffman, who were comparing the spectral signatures of soot particles, generated in the laboratory, with those of interstellar dust particles. Following the publication of the detection of C_{60}, and its proposed structure, Krätschmer and Huffman confirmed, using a combination of mass spectroscopy, ultraviolet and infrared spectroscopy, that their artificially generated soot gave spectra in very good agreement with those predicted for C_{60} with the buckminsterfullerene structure. In particular, they measured four distinct infrared absorption bands—in agreement with group theoretical arguments, which predict only four infrared active (t_{1u}) vibrations for the highly symmetrical C_{60} cluster.

The apparatus used by Krätschmer and Huffman to generate soot involved passing a high current through two graphite rod electrodes in a bell jar, under a low pressure of helium. Using this apparatus, and many variants which were soon manufactured, macroscopic quantities of C_{60}-containing material could now be generated. Krätschmer and Huffman discovered that a component of the soot sample dissolved in benzene to form a red solution. Evaporation of this red solution left behind red crystals, which turned out to consist of a mixture of C_{60} (75%), C_{70} (23%) and higher fullerenes ($\approx 2\%$). X-ray diffraction on these crystals (and on subsequently prepared samples which were of higher purity) confirmed that the C_{60} cluster does indeed have the proposed truncated icosahedral structure and that they pack together in a close-packed face-centred cubic array. The phase diagrams of solid C_{60} and C_{70} (the solids are known as fullerites and constitute new solid forms of elemental carbon) have proven to be complex and very interesting.

Kroto and co-workers at the University of Sussex, subsequently managed to measure the ^{13}C nuclear magnetic resonance (NMR) spectrum of a mixture of C_{60} and C_{70} in benzene solution. They found six distinct resonances, in agreement with theory—since all of the carbon atoms are symmetry-equivalent, the I_h C_{60} molecule shows only one ^{13}C peak, while the lower symmetry (D_{5h}) C_{70} molecule (see Figure 6.3b) has five distinct carbon sites and, therefore, five ^{13}C peaks. The intensity ratios for C_{70} (1:1:1:2:2 or 10:10:10:20:20) agree with the postulated structure.

The spectroscopy and chemistry of buckminsterfullerene and the related fullerenes has been discussed in detail in a number of reviews and books—some of which are listed at the end of this chapter. The discovery of the fullerenes has opened up a new, very rich field of scientific research, as well as introducing a new form of elemental carbon (to add to the familiar diamond and graphite allotropes). The contribution of Kroto, Smalley and Curl in initiating this exciting field was recognized by the joint award of the 1996 Nobel Prize for chemistry.

A number of other fullerenes have also been isolated and structurally characterized. The first higher fullerene to be identified (in 1991) was the chiral (D_2 symmetry) molecule C_{76}, which Ettl identified on the basis of its ^{13}C NMR spectrum—consisting of nineteen lines of equal intensity. Also in 1991, Diederich used high pressure liquid chromatography to separate two isomers of C_{78}, which were then identified by ^{13}C NMR spectroscopy. In fact, three isomers of C_{78} have so far been isolated and two isomers of C_{84} have also been identified. The structures of these higher fullerenes are shown in Figure 6.4.

To date, most fullerenes have been identified by extraction, using an organic solvent (such as benzene, toluene or CS_2) from soot which is generated, as in the Krätschmer and Huffman experiment, using a carbon arc or resistive heating of graphite. There appears to be a decrease in solubility with increasing cluster size, since the raw soot contains even-carbon mass peaks up to well above C_{150}, but the solvent-extracted soots tail off at around C_{96}.

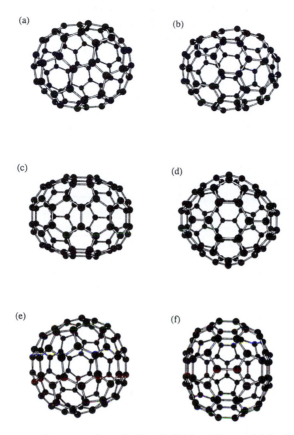

Figure 6.4 Structures of some experimentally identified higher fullerenes: (a) C_{76} (D_2); (b) C_{78} (D_3); (c) C_{78} (C_{2v}); (d) C_{78} (C_{2v}); (e) C_{84} (D_2); (f) C_{84} (D_{2d}).

6.2.2.2 Geometrical Structures of Fullerenes

Carbon clusters lower their electronic energies by lowering the number of unsatisfied valencies (dangling bonds). For small clusters, this is achieved by forming multiple bonds, involving π-bonding. For larger clusters, dangling bonds can be used up in forming rings. Diamond-like (sp^3 hybridized) clusters would have surfaces with lots of dangling bonds, but this number is greatly reduced if the cluster structures are based on graphite, in which the carbon atoms are sp^2 hybridized.

The graphite structure is formed by stacking infinite planar sheets of hexagons. Introduction of a pentagonal defect into a graphitic layer causes it to bend and (as will be shown below), twelve pentagons will cause the graphite sheet to curl up into a ball—leading to the fullerenes.

The fundamental equation governing polyhedral topology is Euler's relation which, for 3-D convex polytopes, takes the form:

$$V + F = E + 2 \tag{6.1}$$

where V, F and E are the total numbers of vertices, faces and edges of the polyhedron respectively. The Euler relationship can be re-expressed as follows:

$$\sum_k f_k - \frac{1}{2}\sum_j (j-2)v_j = 2 \tag{6.2}$$

$$\sum_k v_j - \frac{1}{2}\sum_j (k-2)f_k = 2 \tag{6.3}$$

where there are v_j vertices of connectivity j and f_k faces with k sides and the total number of vertices and faces are given by $V = \Sigma_j v_j$; $F = \Sigma_k f_k$. Equations (6.2) and (6.3) can be combined to give another useful expression of the Euler relationship:

$$\sum_j jv_j = \sum_k kf_k = 2E \tag{6.4}$$

As can be seen from Figures 6.3–6.5, the fullerenes are trivalent polyhedral clusters—that is all carbon atoms are bonded to three others (three-connected)—with pentagonal and hexagonal faces only. For trivalent polyhedra, all vertices have connectivity $j = 3$ and the Euler equation reduces to:

$$\sum_k (6-k)f_k = 12 \tag{6.5}$$

which can be expanded as:

$$3f_3 + 2f_4 + f_5 + (0f_6) - f_7 - 2f_8 ... = 12 \tag{6.6}$$

For fullerenes, which have only pentagonal ($k = 5$) and hexagonal ($k = 6$) faces, Equation (6.6) reduces further to $f_5 = 12$, which explains why all fullerenes have 12 pentagonal faces. The number of hexagonal faces (f_6) is undefined as it has zero weighting in Equation (6.6) and is therefore a variable which gives rise to the family of fullerenes. It is not, however, possible to construct a closed trivalent polyhedron with 12 pentagonal and one hexagonal face, though all other values of f_6 are allowed.

Trivalent polyhedra must always contain an even number of vertices ($N = V$), as the number of edges ($E = 3N/2$) must be an integer. This explains why only even-carbon peaks are observed in the high-mass range of the carbon mass spectrum.

Inserting the total number of faces ($F = 12 + f_6$) and edges into Equations (6.2 and 6.3), we obtain the following relationship between the number of vertices and the number of hexagonal faces of a fullerene:

$$f_6 = \frac{N}{2} - 10 \tag{6.7}$$

$$N = 20 + 2f_6 \tag{6.8}$$

Since f_6 can take any integer value (except 1), fullerenes C_N are possible with N taking any even integer value greater than or equal to 20 (except 22).

These formulae and other geometrical, group theoretical and graph theoretical relationships have been used to develop methods for generating fullerene polyhedra, counting fullerene isomers and predicting their relative stabilities—by deriving rules to relate the electronic and geometric structures of the clusters.

6.2.2.3 Counting Fullerene Isomers

Fowler has stated that for $N \geq 28$ each fullerene has at least two distinct geometrical isomers and that the number of isomers rises very rapidly with N. Fowler and Manolopoulos have enumerated and listed the isomers for a large number of fullerenes, using various constructions.

The Coxeter–Goldberg method was adapted to construct icosahedral deltahedra by triangulating a plane and then wrapping it up into a 3-D polyhedron —generating fullerenes (trivalent polyhedra) as the duals of the deltahedra (triangular-faced polyhedra) thus produced. (The dual operation involves replacing an a-connected vertex in one polyhedron by an a-sided face in its dual and replacing a b-sided face in one polyhedron by a b-connected vertex in its dual.) This led to the generation of two series of icosahedral fullerenes (with $N = 60k$ and $N = 60k+20$, where $k = 0, 1, 2, ...$) which were shown to be closed shell and open shell respectively (see Section 6.2.2.4).

An elegant method for enumerating fullerene isomers is the *spiral algorithm* of Manolopoulos and Fowler, in which the fullerene is conceptually peeled into a single 1-D strip of pentagons and hexagons whose order are recorded as a spiral sequence: 56666656..... etc. The method can be inverted to enable isomer counting by constructing all possible sequences of twelve 5s and ($N/2-10$) 6s and checking for closure and uniqueness. From the spiral sequence, the connectivity pattern can be constructed, which enables the generation of the 3-D fullerene structure. It has been proven that all fullerenes with five-fold or six-fold rotational symmetry axes have at least one spiral: for example icosahedral C_{60} has three spirals. Although it appears that the vast majority of fullerene isomers have at least one spiral, some counter examples have been found: the smallest fullerene without a spiral code is an isomer of C_{380}.

The numbers of fullerene isomers found by Fowler and Manolopoulos for C_{20}–C_{100}, using the spiral algorithm, are listed in Table 6.1. In each case the spirals were started with a pentagon. In the case of C_{100}, it is known, however, that there is at least one isomer which can only be constructed as a spiral starting from a hexagon.

6.2.2.4 Electronic Structures of Fullerenes: Open or Closed Shell?

As well as inspiring experimentalists, the discovery of the fullerenes has also led to a plethora of theoretical studies of carbon cluster molecules, partly due to the aesthetically pleasing highly symmetric nature of buckminsterfullerene (C_{60}), but also because carbon is one of the simplest of elements and one which is amenable to investigation at many levels of theory.

In a fullerene cluster, the cluster framework is formed by overlap of carbon sp^2 hybrids, so that a C_N cluster has $3N/2$ filled bonding and $3N/2$ empty antibonding framework orbitals. The remaining $N p$ orbitals (one on each carbon atom) are oriented radially and would be utilized in bonding to hydrogen in a hydrocarbon cluster. In the following discussion these radial orbitals will be described as 'π-orbitals' as, especially for large clusters, they overlap with each other primarily in a π-fashion. These orbitals interact to generate N π-MOs.

Table 6.1 The number of fullerene isomers N_f and isolated pentagon isomers N_{ip} found by Fowler and Manolopoulos (using the spiral algorithm) for $C_N = C_{20}$–C_{100}, with enantiomers regarded as distinct. As no isolated pentagon isomers are possible for $N < 60$, N_{ip} values are only listed for $N \geq 60$.

N	N_f	N	N_f	N_{ip}
20	1	60	3,532	1
22	0	62	4,670	0
24	1	64	6,796	0
26	1	66	8,825	0
28	3	68	12,501	0
30	3	70	16,091	1
32	10	72	22,142	1
34	9	74	28,232	1
36	23	76	38,016	3
38	30	78	47,868	6
40	66	80	63,416	9
42	80	82	79,023	12
44	162	84	102,684	34
46	209	86	126,973	33
48	374	88	162,793	56
50	507	90	199,128	78
52	835	92	252,082	161
54	1,113	94	306,061	252
56	1,778	96	382,627	349
58	2,344	98	461,020	483
		100	570,602	862

As each carbon atom has one electron in its radial p-orbital, an N-vertex fullerene can only be closed shell (*i.e.* possess no unpaired electrons) if N is even—though this condition is also necessary for the trivalent polyhedron to be geometrically closed. However, N being even is not a sufficient condition for a closed shell fullerene cluster since there remains the possibility of degenerate or pseudo-degenerate frontier orbitals giving rise to open shell configurations for N electrons. An example of this situation occurs for the smallest possible fullerene, dodecahedral C_{20}, with no hexagonal faces. Early experiments indicated that the observed isomers of C_{20} did not include the fullerene and this was attributed to the cluster having an open shell. On the basis of Hückel (and also more sophisticated) calculations, the electronic configuration of 20-(p^π)-electron dodecahedral C_{20} has been determined as $(a_{1g})^2(t_{1u})^6(h_g)^{10}(g_u)^2$. Interestingly, the photoelectron spectroscopy of mass selected C_{20}^- ions (generated by the replacement of the hydrogen atoms in dodecahedrane, $C_{20}H_{20}$, followed by debromination and electron attachment) have been interpreted in terms of the dodecahedral fullerene structure, even though this species would still be open shell under I_h symmetry and presumably should undergo a symmetry-lowering Jahn–Teller distortion.

The necessary criteria for (properly) *closed shell* N-vertex fullerenes (with N even) are:

(i) the spectrum of π-orbitals has $N/2$ low-lying (bonding) and $N/2$ high-lying (antibonding) components,

(ii) orbitals $N/2$ and $(N/2)+1$ are not degenerate or pseudo-degenerate (i.e. there is a significant HOMO–LUMO separation).

Fowler has also identified *pseudo-closed shell* fullerenes (where there is a HOMO–LUMO gap, but the LUMO is weakly bonding and low-lying) and *meta-closed shell* fullerenes (where there is a HOMO–LUMO gap, but the HOMO is antibonding and relatively high-lying). These possibilities, along with the open shell isomers, will generally be less stable than properly closed shell isomers.

A number of methods for predicting and generating properly closed shell fullerenes have been described. The most significant work in this area is also due to Fowler and Manolopoulos, who have used group and graph theoretical methods, such as the *leapfrog* construction, underpinned by Hückel, semi-empirical or *ab initio* MO calculations. For an arbitrary fullerene geometry, however, a properly closed electronic shell is an exception rather than the rule: of the first 1.5 million fullerene isomers only 17 are properly closed shell!

6.2.2.5 Relative Stabilities of Fullerene Isomers

The relative stabilities of fullerene isomers are governed by two factors: the energies of the $N/2$ occupied π-MOs; and the strain energy of the sp^2 framework. The total radial orbital energy is related to the open or closed shell nature of the fullerene, since properly closed shell clusters have large HOMO–LUMO gaps (Δ) and, therefore, the sum of the one-electron π-electron energies will be lower. While small Δ values, or open shells, are usually associated with low stability, a large Δ does not always imply greater stability, due to the greater importance of the steric preferences of the sp^2 framework.

Considering the strain energy of the sp^2 backbone, the main structural feature governing isomer stability is the *isolated pentagon rule*, which states that for a given nuclearity the relative energies of fullerene isomers increase with increasing numbers of adjacent pentagons. The isolated pentagon rule was first proposed by Kroto, based on experimental mass spectral results, and has subsequently been supported by theoretical calculations. Indeed, to date, all of the experimentally characterized fullerenes (with the exception of the dodecahedral C_{20}) have isolated pentagons, though Fowler and Manolopoulos have shown that isolated pentagons only occur in a small percentage of fullerene isomers. The buckminsterfullerene isomer of C_{60} is the smallest fullerene with isolated pentagons. Although there are no isolated-pentagon isomers from C_{62} to C_{68}, it has been shown (see Table 6.1) that for $N \geq 70$ there is always at least one isomer with isolated pentagons. Given a choice of isomers with the same number of adjacent pentagons, Raghavachari has postulated that the most stable isomers will have the most even distribution of hexagon environments.

These stability rules have been extended to include non-classical fullerene isomers with square and/or heptagonal faces, in addition to the usual pentagons and hexagons. For fullerenes with heptagonal faces, Equation (6.6) indicates that there has to be an additional pentagonal face for every heptagon added (since $f_5 - f_7 = 12$). In the case of C_{62}, an isomer with one heptagon is actually predicted to be more stable than any classical fullerene isomer.

6.2.2.6 Fullerene Isomerization

The huge number of isomers possible for fullerenes of a particular size, C_N, leads to an interesting conundrum. One would expect that, if cluster growth proceeded via a random self-assembly process, a very large number of isomers would be generated. It is known, however, that the products of fullerene synthesis generally contain no more than a few (or even one) isomers for each value of N. It seems likely, therefore, that at the high temperatures pertaining to fullerene formation, the initially formed cages have sufficient energy to rearrange, yielding a few, thermodynamically favoured (low energy) isomers. The difference between alternative fullerene isomers can be described in terms of their different arrangement of pentagons, so this isomerization must be associated with the effective motion of the pentagons over the surface of the cluster cage.

A number of mechanisms have been proposed for fullerene isomerization, the most popular being the *Stone–Wales* or *pyracyclene rearrangement*. The Stone-Wales (SW) rearrangement involves a local $\pm 90°$ rotation of the central C_2 fragment in a pyracylene patch (where two pentagons and two hexagons meet) on a fullerene. As shown in Figure 6.5, this process leads to the interconversion of the two pentagons and the two hexagons (the patch has effectively been rotated by 90°) with the product of the rearrangement also being a fullerene. Fowler and Manolopolous have mapped out isomerization pathways for a number of fullerenes, based on SW rearrangement steps.

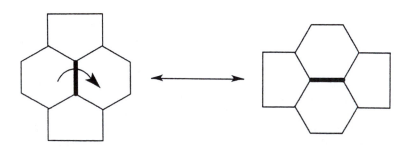

Figure 6.5 The Stone–Wales rearrangement leads to the interconversion of fullerene isomers.

The SW transformation is a high energy process, with *ab initio* calculations predicting activation barriers in excess of 500 kJ mol^{-1} for the concerted SW rearrangement. Calculations have shown that in some cases lower energy barriers may result from sequential 1,2-shifts, proceeding via a non-fullerene intermediate with an sp^3–sp hybridized C_2 fragment.

6.2.2.7 Fullerene Growth and Fragmentation

The mechanism of fullerene growth is still not well understood and a number of questions remain unanswered—such as: why is buckminsterfullerene (icosahedral C_{60}) so abundant in the products of condensation of carbon vapour? It is evident that the stable fullerenes, such as C_{60}, must form via accretion of smaller fragments, as the simultaneous condensation of (say) 60 carbon atoms is extremely improbable.

Two distinct mechanisms have been proposed for the formation of C_{60} from small polycyclic fragments. In the *pentagon road* model, the minimization of dangling bonds in growing graphite-like sheets results in pentagon 'defects' which cause the sheets to curl up. Rearrangements (SW or otherwise) remove destabilizing adjacent pentagons and fullerenes form when the deformed graphite-like sheet closes up. If the pentagons appear at the wrong time or in the wrong place, further cluster growth may occur—perhaps leading to spiral multilayer particles. Such an *icospiral* nucleation scheme has been proposed as a possible mechanism for the formation of macroscopic soot particles under conditions of high carbon density.

In the *fullerene road* model, larger fullerenes grow from smaller (completely formed) fullerenes, with around 40 atoms, by the insertion of C_2 or other small molecules, with annealing (via SW pathways etc.) resulting in the removal of pentagon adjacencies. Intense laser irradiation of the higher fullerenes has been found to cause fragmentation—probably by loss of C_2 fragments—as a series of peaks, corresponding to smaller (even-membered) fullerenes are observed, down to C_{32}. Whereas the only detectable products of C_{34} fragmentation are C_{32} and C_2, C_{32} itself fragments to give a mixture of carbon chains—as for this smaller fullerene loss of C_2 followed by cage reclosure may lead to excessive strain energy. A mechanism for the loss of a C_2 fragment from fullerenes, while conserving the number (twelve) of pentagons, involves the loss of a C_2 unit which is shared by two adjacent pentagons. (This high energy isomer can be considered as a fragmentation intermediate, which may be generated from a more stable isomer by one or more SW rearrangements.) This process results in the conversion of two hexagons into pentagons, while the original adjacent pentagons become a single hexagon. The excision of C_2 therefore results in the overall loss of one hexagon. The reverse process can be envisaged as a mechanism for C_2 insertion into a fullerene.

Interestingly, irradiation of solid C_{60} and C_{70} leads to the production of a range of larger, as well as smaller, fullerenes. Presumably fragments lost by an irradiated fullerene C_N can be inserted into another cluster, to generate species $C_{N\pm2}$, $C_{N\pm4}$, etc. when the clusters are in close proximity to each other, in the solid state.

6.2.2.8 Fullerene Reactivity

Because they can be produced in large quantities, most reactivity studies have been performed on C_{60} and C_{70}. It was at first assumed that C_{60} would be a relatively unreactive molecule, since it has a very high number (approximately 12,500) of resonance structures (ways in which double and single bonds can be permuted) which was believed to impart 'superaromatic' character to the cluster. The reactivity of a superaromatic molecule would be low as the transition state would involve the disruption of the electron delocalization over the cluster framework.

In fact, C_{60} behaves more like a superalkene than a superaromatic molecule. Neutron and X-ray diffraction studies of solid crystalline C_{60} reveal bond lengths of 1.46 Å (within the pentagons) and 1.39 Å (radiating from the pentagons). Thus, a localized bonding picture of C–C single and double bonds is appropriate, leading to the Kekulé structure of alternating single and double bonds within the hexagons (which are said to be 'benzenoid') as shown in Figure 6.6. It should be noted that the earlier MO calculations of Bochvar and G'alpern had correctly predicted the localized nature of the bonding in C_{60}. This bond localization (or 'fixation') occurs because double bonds within pentagons give rise to high strain energies. If fullerene isomers are drawn such that there are no double bonds in the pentagonal rings and each atom has a valency of four (i.e. has four bonds to it) then the lowest energy isomers tend to be those with the greatest number of benzenoid rings.

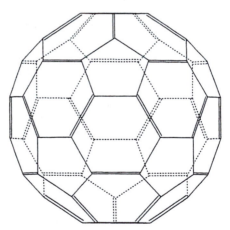

Figure 6.6 The Kekulé (superalkene) structure of C_{60}.

As a superalkene, C_{60} readily undergoes addition reactions across the double bonds. While normal alkenes (such as ethene) are electron rich and undergo electrophilic addition, C_{60} acts as a more electron-deficient species (perhaps due to partial delocalization of π-electron density into the pentagonal rings) and nucleophilic addition (of electron donors such as amines) is also known.

Other examples of addition reactions include the addition of oxygen to form epoxides ($C_{60}O$), the addition of carbenes, osmium tetroxide and other transition metal fragments. Many of these addition complexes have now been structurally characterized by X-ray crystallography. C_{60} can be reduced in liquid ammonia (the Birch reduction) to generated hydrogenated products such as $C_{60}H_{36}$ and possibly $C_{60}H_{56}$. Such extensive addition will result in products with primarily C–C single bonds, as the electrons formerly used in π-bonding are now used in forming C–H bonds, with a corresponding change in hybridization of the carbon atoms from sp^2 to sp^3.

C_{60} can be extensively fluorinated—as far as the fully saturated molecule $C_{60}F_{60}$—though the more highly fluorinated clusters eliminate F_2 on standing. The fluorines can be replaced by other negatively charged or electronegative species, by nucleophilic substitution, in contrast to perfluoro-alkanes such as polytetrafluoroethylene (PTFE or Teflon®). Reaction with the heavier halogens leads to species such as $C_{60}Cl_{12}$, $C_{60}Cl_{24}$, $C_{60}Br_6$, $C_{60}Br_8$ and compounds such as $C_{60}Br_{28}$ which possess both cage-bound and unbound bromine and can better be written as a solvated cluster: $C_{60}Br_{24}\cdot2Br_2$. Phenly and alkyl groups can also be attached to the cage, either via electrophilic/nucleophilic addition or in free-radical reactions.

6.2.2.9 Metal–Fullerene Complexes

In the early days of fullerene studies, the Rice–Sussex team managed to incorporate metals (such as potassium, lanthanum and uranium) into the C_{60} cage by laser vaporization of a graphite sample that had previously been immersed in a solution of the appropriate metal salt. These metal atoms, whose presence was detected by mass spectroscopy, could not be removed by laser irradiation and it was proposed that the metal atom was inside the cluster cage ('endohedral'), as shown in Figure 6.7a, with the endohedral complex being denoted $M@C_{60}$. Intense laser irradiation of these species resulted in sequential C_2 loss from the cluster, with the number of carbon atoms in the final endohedral metallofullerene product $M@C_N$ depending on the ionic radius of the metal atom. Thus, the metal has effectively been 'shrink-wrapped'. The smallest endohedral potassium complex is $K@C_{44}^+$, while the larger caesium ion cannot be accommodated inside a fullerene with fewer than 48 carbon atoms. In contrast, there is evidence that fullerenes with more than 88 carbon atoms may be able to accommodate up to three lanthanum atoms.

The metals which have been encapsulated within fullerene cages have been the more electropositive metals, which undergo an energetically favourable (metal to cage charge transfer) interaction with the cluster. These elements are also known to form strong charge-transfer intercalation compounds with graphite. Interestingly, the species K_3C_{60} has been identified, which has one (endohedral) potassium inside the cage, with the other two being attached to the outside of the cage ('exohedral')—so that the complex may better be represented as $K_2(K@C_{60})$.

The lowest energy isomer for the fullerene C_{28} is believed to possess tetrahedral (T_d) symmetry, as shown in Figure 6.7b. The structure has 4 atoms, arranged at the vertices of a tetrahedron, which are each at the junction of three pentagons. *Ab initio* calculations have shown that the electronic ground state of C_{28} is $^5A_2$, with an unpaired electron associated with each of these atoms.

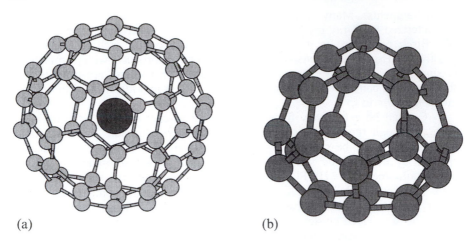

(a)   (b)

Figure 6.7 (a) Representation of an endohedral metallofullerene. (b) The tetrahedral structure of C_{28},
which can form endohedral and exohedral complexes.

These calculations also indicate that the C_{28} cluster should behave as a hollow
'superatom' with a valency of four, which is capable of bonding (via the
tetrahedral vertices) both inside and outside the cluster. This has been confirmed
by the formation of endohedral metallofullerenes $M@C_{28}$ (M = U, Ti, Zr, Hf) with
metals which readily adopt the +4 oxidation state. Exohedral derivatives such as
$C_{28}H_4$ have also been predicted to have high stability.

There is growing experimental evidence that the presence of endohedral
metal atoms can change the order of stability of fullerene isomers. In the case of
C_{80}, for example, there are nine possible isolated pentagon isomers (see Table
6.1—though this includes two pairs of enantiomers). Two of these isomers (with
D_2 and D_{5d} symmetry) have been isolated from arc-processed soot. The
endohedral metallofullerene $La_2@C_{80}$ has been generated in reasonable
abundance and ^{13}C and ^{139}La NMR studies indicate rapid circular motion of two
La^{3+} ions inside a C_{80}^{6-} cage, which has I_h symmetry. Theoretical calculations
confirm that the I_h isomer is closed shell and should be most stable for C_{80}^{6-}, while
the D_2 and D_{5d} isomers are closed shell and lower in energy for neutral C_{80}.

By co-evaporation of metal atoms with pre-formed C_{60}, Martin and co-
workers have shown that C_{60} can be fully coated by metals such as calcium and
barium. In the case of calcium, a particularly strong peak was observed in the
mass spectrum corresponding to $Ca_{32}C_{60}$—perhaps corresponding to one calcium
atom sitting over each of the twelve pentagonal and twenty hexagonal faces. A
corresponding intense peak is observed for $Ca_{37}C_{70}$ which is consistent with this
hypothesis, since C_{70} has 37 faces. Further analysis of the mass spectra of M_xC_{60}
(M = Ca, Ba) reveals magic number peaks at x = 104, 236 and 448—see Figure
6.8. These numbers are consistent with icosahedral packing of the metal atoms,
with the total number (N_M) of metal atoms in K shells around the C_{60} core being
given by:

$$N_M(K) = \frac{1}{3}\left(10K^3 + 30K^2 + 56K\right) \tag{6.9}$$

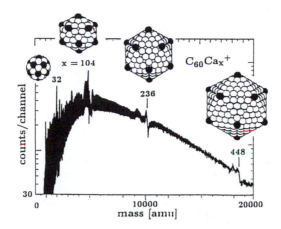

Figure 6.8 Mass spectrum of calcium coated C_{60} clusters. Magic numbers correspond to complete icosahedral layers of metal atoms.

In exo- and endohedral metallofullerenes, the metal atoms are generally positively charged, so the fullerene carries a negative charge. The link with graphite intercalation compounds is strengthened if one considers the alkali metal fullerides—solid compounds with compositions such as M_3C_{60}, M_6C_{60} (M = K, Rb, Cs, etc.) where the positively charged alkali metal ions sit in interstices between the negatively charged (C_{60}^{3-} or C_{60}^{6-}) clusters, which are arranged in a close-packed fashion. The metal ions are thus exohedral and these fulleride compounds may be regarded as doped fullerite solids. Other doping species include iodine. The alkali metal fullerides have attracted much attention since some of them possess superconducting properties, albeit at fairly low temperatures—thus K_3C_{60} becomes a superconductor below 18 K. The highest superconducting temperature (T_c) so far obtained for a fulleride is 33 K for $(Rb,Cs)_3C_{60}$. In fact, this compound has the highest T_c of any molecule-based material.

6.2.2.10 Prehistoric Extra-terrestrial Fullerenes?

In a fascinating recent development, the ability of C_{60} and other fullerenes to encapsulate atoms of other elements has been exploited to gain more information on the Permian–Triassic mass extinction (approximately 250 million years ago) which is believed to have been caused by an asteroid colliding with the Earth. Becker and colleagues (at the University of Washington in Seattle) have detected fullerenes (C_{60}–C_{200}) in sediments at the Permian–Triassic geological boundary. The isotopic ratios measured for rare gases (such as He and Ar) trapped in these fullerenes are similar to those measured in a class of meteorites, known as carbonaceous chondrites, which form around carbon stars (e.g. the He is present mostly as ^{3}He, rather than ^{4}He). It appears, therefore, that fullerenes can be formed in space, probably around carbon stars, and that they can be used as tracers for events dating back millions of years!

6.2.2.11 Large Fullerene-like Clusters

(a) Large fullerenes. As fullerenes grow larger they tend to become more polyhedral and facetted, rather than spherical, with large almost planar graphite-like ('graphene') sheets of hexagons and all of the strain associated with the twelve pentagonal corners. Thus is confirmed by theoretical calculations on large icosahedral fullerenes, such as C_{240} and C_{540}. Simulations of transmission electron microscope (TEM) images of large carbon particles have shown that whether they appear spherical or facetted depends on their orientation relative to the viewing axis.

(b) Hyperfullerenes. Under intense electron beam irradiation, carbon soot particles and buckytubes (see below) curl up into pseudo-spherical carbon particles, which have been found (using TEM) to consist of multiple concentric shells of graphite-like material. These structures have been interpreted in terms of a set of successively larger fullerenes which are nested inside one another— rather like a Russian doll. These clusters have been named hyperfullerenes or 'buckyonions' because of their shell-like nature. As mentioned above, if fullerene closure fails, then large spiral particles may result, though these should be unstable with respect to the buckyonions.

(c) Fullerene polymers. Irradiation of solid C_{60} with visible or UV light leads to the formation of linear chain polymers, where neighbouring molecules are linked by four-membered rings. The mechanism is believed to occur via a concerted electrocyclic reaction. Polymerization can also be induced at high hydrostatic pressures (5 GPa) at temperatures of 500–800°C. A variety of doubly and singly linked polymers are now known, including alkali metal doped polymers which display metallic conduction in one dimension.

(d) Clusters of fullerenes. As fullerenes, such as C_{60}, are known to form close packed arrays in the solid state, it would be expected that they would also cluster together in the vapour phase. Martin and co-workers have measured the mass spectra of clusters of C_{60}, which may be written as $(C_{60})_N$, produced in a molecular beam. As can be seen from Figure 6.9, these 'clusters of clusters' have mass spectra which are reminiscent of those measured for argon and other rare gas clusters—with geometric shell magic numbers at $N = 13$ and 55 and sub-shell magic numbers at $N = 19, 23, 43, 46$ and 49. Analysis of the shell and sub-shell structure has indicated that, like the rare gas clusters, these clusters have an icosahedral packing of C_{60} molecules. This has been confirmed by Wales, who has modelled the clusters using empirical potentials.

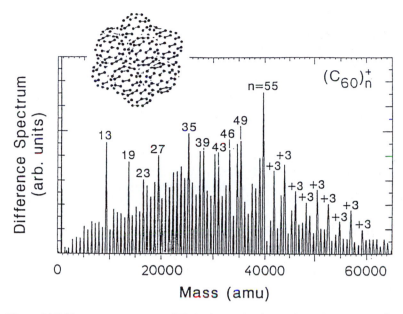

Figure 6.9 Difference mass spectrum of $(C_{60})_N$ clusters showing magic numbers corresponding to icosahedral packing.

6.2.2.12 Substituted Fullerenes

A number of fullerenes have been synthesized with one or more carbon atoms replaced by another element, such as boron or nitrogen, which readily adopts a trivalent coordination. Examples based on C_{60} include $C_{59}B$ and $C_{59}N$ (azafullerene). Both of these species are radicals, being isoelectronic with C_{60}^+ and C_{60}^-, respectively. Indeed the $C_{59}N$ cluster readily dimerizes to form $(C_{59}N)_2$, which has two cages linked by a C–C (rather than a N–N) bond. In the $C_{59}N$ radical, the nitrogen atoms are electronically saturated, each possessing an exo-lone pair of electrons, while the unpaired electron resides on a neighbouring carbon atom.

6.2.2.13 Applications of Fullerenes

Possible applications for fullerenes and compounds derived from fullerenes are virtually endless. Some of the suggested applications include the use of encapsulated radioactive radon atoms for labelling, medical imaging and cancer radio-chemotherapy—perhaps via functionalizing the outside of the fullerene and attaching it to a tumour-targetting antibody. Surface-functionalized fullerenes could also be used as catalysts and thin films of fullerenes may find applications in optical and electronic devices.

6.2.3 Carbon Nanotubes

In 1991, Iijima (at the NEC company in Japan) discovered long filaments of rolled up graphite-like carbon 'nanotubes' growing on the cathode in arc discharge experiments between carbon electrodes under a low pressure of helium. These nanotubes had lengths ranging between 100 nm and several μm, and consisted of from two to twenty concentric cylinders. The internal diameters of the tubes lay in the range 1–20 nm. In later experiments, it was found that single layer nanotubes (also known as 'bucky tubes') form on the walls of the arc chambers. These single-walled tubes have internal diameters of 1–5 nm. As shown in Figure 6.10, the walls of the nanotubes consist of graphene sheets of hexagons, while the ends are capped by hemispherical fullerene-like groups, possessing pentagons as well as hexagons.

Nanotubes have a number of potential applications. The strength of the binding in these structures means that nanotubes may eventually replace traditional (macroscopic) carbon fibres in lightweight, high strength composite materials. Iijima and co-workers demonstrated that heating nanotubes to 400°C in molten lead results in Pb atoms being sucked into the nanotubes by capillary action. A number of metals and metal salts have now been incorporated into nanotubes, opening up the possibility of performing shape- and size-selective catalysis inside nanotubes, in an analogous way to catalysis in zeolite pores. Incorporation of silver into nanotubes leads to metallic nanowires, of length 20–100 nm, which may ultimately find use in nanoscale electronic devices. The ability to insert biological molecules into nanotubes may lead to their use in biomedical applications.

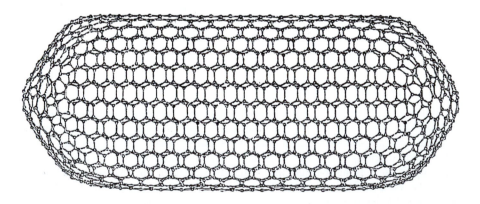

Figure 6.10 Schematic representation of a closed nanotube.

6.3 SILICON CLUSTERS

6.3.1 Mass Spectra of Silicon Clusters

The mass spectra of silicon clusters (Si_N^+) were first reported by the groups of Bloomfield and Smalley around 1985. The clusters were generated by laser flash evaporation, quenched in a carrier gas and then cooled by supersonic expansion. The mass spectrum (reproduced in Figure 6.11) reveals relatively high abundances for Si_N^+, with N = 6, 10, 16 and 32. It is noticeable, however, that unlike for metal and rare gas clusters, a large peak for N atoms is not followed by a minimum for $N+1$ atoms. This has been interpreted in terms of laser vaporization not leading to cluster annealing in the case of semiconductor clusters. In contrast to metal and rare gas clusters, the binding of an additional silicon atom to a magic number core is not much greater than the binding energies within the magic number cluster.

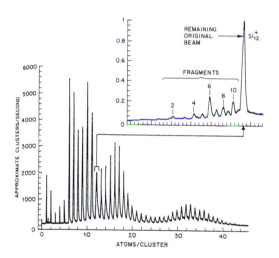

Figure 6.11 Mass spectrum of silicon clusters and (inset) the photofragmentation spectrum of Si_{12}^+.

6.3.2 Photofragmentation Studies

In the experiments of Bloomfield and Smalley, mass selected silicon clusters were selectively photofragmented, using laser radiation. Monomers were found to be unimportant in cluster fragmentation (again unlike alkali metal and rare gas clusters, which readily evaporate single atoms), which is consistent with the low abundance of monomers in the equilibrium vapour of these elements. Instead, silicon clusters tend to fragment into smaller clusters:

$$\text{Si}_N^+ + h\nu \rightarrow \text{Si}_{N-M}^{(+)} + \text{Si}_M^{(+)}$$

where the parentheses round the plus signs indicate that the positive charge may reside on either fragment. Further fragmentation is possible. For clusters of around ten atoms, the fragments are often of approximately equal size and symmetric fragmentation (or fission) is encountered, even for larger clusters:

$$\text{Si}_{12}^+ + h\nu \rightarrow \text{Si}_6^+ + \text{Si}_6$$
$$\text{Si}_{14}^+ + h\nu \rightarrow \text{Si}_7^+ + \text{Si}_7$$
$$\text{Si}_{20}^+ + h\nu \rightarrow \text{Si}_{10}^+ + \text{Si}_{10}$$

For clusters with $N \geq 12$, fragmentation generally involves loss of $\text{Si}_M^{(+)}$ fragments, where M is in the range 6–11, with Si_6 and Si_{10} having high abundance in the photofragmentation patterns—as can be seen in Figure 6.11 for the photofragmentation of Si_{12}^+. For larger clusters ($N > 30$) the main fragmentation channel involves loss of 6–11-atoms, though there is also evidence for loss of a single neutral silicon atom. Thus, the photofragmentation pattern for Si_{60}^+ shows large peaks for $\text{Si}_6^+ - \text{Si}_{11}^+$, and a small peak for Si_{59}^+.

The reduced tendency for monomer evaporation arises because the covalently bonded silicon clusters are more rigid than the (liquid-like) alkali metal or rare gas clusters. Cluster relaxation is slow and there is no evaporation, so cooling is achieved by collisions with the carrier gas. In this way, silicon (and other semiconductor) clusters behave like transition metal clusters (e.g. Fe_N, where there is also slow relaxation and little monomer evaporation).

Smalley and co-workers have found that small differences in experimental conditions—in particular the carrier gas—can lead to large changes in the size distribution of silicon clusters. Phillips has attributed this sensitivity to the large barriers to rearrangement in covalent network materials which are responsible for their tendency to form amorphous or glassy solids. Thus, as mentioned above, the silicon clusters are not annealed in the cluster source.

6.3.3 Cluster Mobility Studies

In 1990, Jarrold and co-workers measured the mobilities of mass-selected charged silicon clusters, Si_N^+, with N in the range 10–60, using an ion drift tube (see Section 1.3.5.2). For silicon, 50 μs pulses of clusters were injected into the drift tube and subjected to a weak electric field (3.28–22.97 V cm^{-1}). The cluster ions were injected with an energy of 50 eV, so as to overcome the pressure (approximately 10 Torr) of the helium buffer gas. The arrival time distribution of cluster ions was measured at the detector at 10 μs intervals.

The arrival time distributions for Si_N^+, with $N = 10$–23 (Figure 6.12), show a single peak in each case, indicating that either only one isomer is present for each nuclearity or, if more than one isomer is present, that they must have very similar topologies. In the range $N = 24$–35, however, two components can clearly be resolved in the arrival time distributions, indicating the presence of two isomers with significantly different topologies, and hence quite different mobilities.

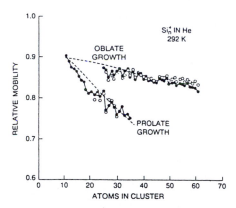

Figure 6.12 Plot of the relative mobilities (measured mobility divided by hard sphere mobility) of charged Si clusters in He. The data correspond to injection energies of 50 eV (filled circles) and 130 eV (empty circles). Dashed lines indicate predicted relative mobilities for prolate and oblate clusters.

Analysis of the arrival time data led Jarrold to conclude that for $N = 10$–23, a single isomer, with a prolate ('sausage-shaped'—i.e. with one long axis and two shorter axes) topology is present for each cluster species Si_N^+. As the clusters get larger, the relative mobility decreases, giving a series of points labelled 'prolate growth' in Figure 6.12. At $N = 24$, a second series of isomers appear, which have greater mobilities than the prolate series. This second series has been attributed to oblate (i.e. with two long axes and one shorter axis) cluster topologies. For a given nuclearity, and hence volume (if the cluster density is assumed to be constant), an oblate cluster isomer (which is more spherical than a prolate isomer) has a smaller effective collision cross section as it rotates (than does the prolate isomer) and, thus, has a higher mobility.

As shown in Figure 6.12, the mobilities of the oblate series of isomers (labelled 'oblate growth') decreases more slowly than the prolate series, which is again consistent with their assignment as oblate. Beyond $N = 35$, it is apparent that the prolate series has terminated, so that only a single pseudo-spherical oblate isomer is present.

Jarrold also performed annealing studies on the silicon clusters, by increasing the injection energy to 130 eV, so that collisions between the cluster ions and the helium carrier gas impart greater internal energy to the cluster, which in turn can lead to cluster reconstruction. For Si_{32}^+, for example, at the higher injection energy the slow moving component (the prolate isomer) which was observed in the 50 eV experiment disappears, as it anneals into the more spherical oblate isomer. Figure 6.12 shows the effect of annealing for clusters in the range $N = 20$–60. For the smaller clusters, there is not much change, though the oblate isomer of Si_{24}^+ disappears—perhaps indicating that (as oblate isomers have just started to appear at $N = 24$) the oblate isomer is thermodynamically unstable with respect to the prolate form. In the region $N = 25$–29, both isomer types are still observed after annealing—indicating roughly equal stabilities. The post-annealing relative (prolate/oblate) isomer abundances vary significantly, however. For Si_{36}^+, the area of the prolate isomer peak is greater than that of the oblate peak, indicating a greater abundance of the prolate isomer after annealing. By contrast, for Si_{29}^+ the oblate isomer peak is much more intense than the prolate peak. Jarrold has estimated the activation energy for the prolate to oblate interconversion to be 1.38 eV for Si_{29}^+. From $N = 30$–35, annealing results in the complete loss of the prolate isomer—indicating that the prolate isomers are thermodynamically unstable with respect to the oblate isomers in this size regime. This is consistent with the absence of prolate isomers for unannealed clusters with $N > 35$.

6.3.4 Reactivity of Si Clusters

Jarrold and co-workers have also studied the reactivity of silicon clusters in their drift tube experiments. Analysis of the results of reactive collisions with ethene (C_2H_4), leads to the conclusion that there may be two or three distinct isomers of Si_{26}^+ and at least four isomers of Si_{25}^+. Non-reactive collisions with small molecules, such as ethene, oxygen and ammonia, can also lead to cluster annealing.

It should be noted that the most thermodynamically stable isomer is not always the most unreactive—thus in reactive collisions of Si_{29}^+ with ethene, the oblate isomer (which, after conventional annealing, is more abundant than the prolate isomer) is almost entirely removed. Reactivity patterns are not simple, however, as can be shown by the relative reactivity of oblate and prolate isomers of Si_N^+ with O_2. For Si_{26}^+, Si_{27}^+ and Si_{30}^+, the oblate isomer is roughly twice as reactive towards O_2 as the prolate isomer, while the reverse is true for Si_{25}^+, Si_{28}^+ and Si_{29}^+. For Si_{24}^+, the reactivities of the prolate and oblate isomers are approximately equal.

The nature of the reaction with oxygen also changes with cluster size. Thus, for $N \lesssim 29$, the reaction primarily involves etching of the cluster, via loss of two SiO molecules:

$$Si_N^+ + O_2 \rightarrow Si_{N-2}^+ + 2SiO$$

This is an exothermic reaction (by 165–250 kJ mol^{-1}) but the reaction is very slow (particularly for the larger clusters) and depends strongly on cluster size. For larger clusters ($N > 36$) the oxygen sticks to the surface cluster, probably adsorbing dissociatively:

$$Si_N^+ + O_2 \rightarrow [Si_N O_2]^+$$

The reaction shows little size-dependence. The different types of reaction and their relative probabilities are indicated in Table 6.2.

Table 6.2 Comparison of reaction mode and reaction probability for silicon clusters of varying sizes.

Cluster	Reaction type	Reaction probability
Si_{10}^+	Etching	0.02
Si_{13}^+	Etching	0.00004
Si_{40-65}^+	Adsorption	0.0012
Si_N^+ (on graphite)	Adsorption	0.001–0.002
Bulk Si surface	Adsorption	0.1

Measurements have been also been made of the reactivity of neutral silicon clusters which have been deposited onto a graphite substrate (or 'support'). For all cluster sizes, O_2 is found to adsorb onto the clusters, and no etching is observed. Little variation of reactivity is observed with changing cluster size (see Table 6.2). It is therefore apparent that supported clusters exhibit markedly different chemical reactivity compared to free gas-phase clusters. The strong interaction between the cluster and the supporting substrate (the 'support effect') can lead to significant geometric and electronic relaxation of the cluster. In particular, the good thermal contact between the cluster and the substrate allows energy to flow from the cluster to the substrate, thereby offering alternative relaxation channels to the cluster.

Finally, it should be noted that all of the silicon clusters studied are significantly less reactive than bulk silicon surfaces. For example, the sticking probability of O_2 (dissociative adsorption) on the surface of bulk surface (0.1) is two orders of magnitude greater than for the largest silicon clusters studied. This may indicate that the cluster structures are such that there are few active sites for adsorption and dissociation.

Smalley and co-workers investigated the reaction of silicon cluster ions (Si_N^+, with $N \le 65$) with ammonia, detecting products $[Si_M(NH_y)]^+$. In Smalley's experiment, the silicon clusters were not mass-selected, so mass spectral abundances were compared before and after exposure to ammonia. Smalley found that certain clusters (Si_{33}^+, Si_{39}^+ and Si_{45}^+) are relatively unreactive towards NH_3, while others ($Si_{42}^+ - Si_{44}^+$ and Si_{46}^+) exhibit high reactivity. For $N > 47$, reactivity increases slowly and smoothly with increasing cluster nuclearity, as the surface area of the clusters increases.

6.3.5 Structures of Si Clusters

There have been numerous attempts to rationalize the experimental results discussed in the preceding sections and to use these results to infer the structures and growth patterns of the silicon clusters. There have also been a number of theoretical studies of silicon clusters, using calculation methods of varying degrees of sophistication.

In order to interpret the mass spectra and photofragmentation patterns of silicon clusters (where Si_6^+ and Si_{10}^+ are present in high abundance), Phillips proposed that larger silicon clusters are composed of distorted octahedral Si_6 and adamantane (a diamond fragment) Si_{10} fragments. He also postulated diamond-like fragments for Si_{14}^+ and Si_{16}^+.

High level DFT calculations have been performed on silicon clusters with up to ten atoms by Raghavachari. The lowest energy structures calculated for Si_2–Si_{10}, which are reproduced in Figure 6.13, indicate that, in this size regime, the structures are quite distinct from crystalline diamond-like structures. This structural rearrangement is driven by the requirement to lower the energy of the large number of surface dangling bonds which would otherwise result. There is, therefore, a tendency towards higher coordination numbers and smaller rings (three- and four-membered rings) than in bulk silicon (where each atom is four-coordinated and all rings are six-membered) and, therefore, to clusters which are more condensed than the diamond structure.

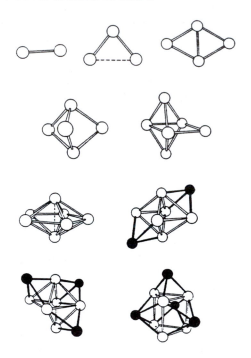

Figure 6.13 Structures of neutral silicon clusters calculated using Density Functional Theory. Capping atoms are indicated by shading.

It should be noted that the calculated lowest energy structure for Si_{10} is the high density tetracapped trigonal prism, rather than the adamantane structure proposed by Phillips. (As will be discussed later, however, such diamond-like fragments are believed to be among the more stable geometries for hydrogenated silicon clusters.) The reduced tendency for silicon to participate in π-bonding, compared with carbon, explains why long silicon chains and rings are not observed. In fact, the structures of the smaller silicon clusters resemble those of lithium or sodium clusters more closely than carbon clusters. The relative weakness of Si–Si π-bonds (compared with the formation of a greater number of σ-bonds) also rules out hollow, π-bonded, fullerene-like structures for the larger silicon clusters. Semi-empirical calculations confirm that the icosahedral (buckminsterfullerene) structure has low stability for Si_{60}.

Raghavachari's calculations showed little difference in structures for neutral and cationic silicon clusters, since on going from Si_N to Si_N^+, the electron is generally removed from a molecular orbital with non-bonding or weakly bonding character. The calculated binding energies per atom were found to give local maxima at $N = 6$ and $N = 10$, which is consistent with the experimental photofragmentation studies. In fact, the photofragmantation patterns of clusters with up to twenty atoms were found to correlate well with the calculated stabilities of the possible fragments.

On the basis of tight binding MO calculations, George and co-workers have postulated structures for clusters in the range $Si_{30}^+ - Si_{45}^+$ which are based on stacks of planar hexagons, with a number of capping atoms (for example, Si_{37}^+ and Si_{39}^+ consist of six hexagons and a one- or three-atom capping unit). These structures were used to explain the photofragmentation and reactivity trends for silicon clusters in this size range. These workers also proposed a structure for Si_{60}^+, to account for its principal fragmentation product Si_{10}^+, which is composed of six stacked napthalene-like Si_{10} units (two edge-sharing hexagons).

In order to explain his cluster mobility measurements, as well as the photofragmentation mass spectra, Jarrold has proposed that the prolate isomers for Si_N^+ (up to $N = 35$) are built up of stable tricapped and tetracapped trigonal prisms—thus Si_{28}^+ would be built up as a stack of two tricapped trigonal prisms and a tetracapped trigonal prism. The more spherical (oblate) geometries which predominate above $N = 35$, cannot be decomposed into such fragments.

Kaxiras generated structures for pseudo-spherical silicon clusters based on four-coordinated interior (bulk-like) atoms and three-connected surface atoms—thus his model for Si_{29} has five four-coordinate Si atoms and twenty-four three-coordinate surface atoms. He suggested that the reduced reactivity of clusters such as Si_{33}^+ and Si_{45}^+, towards small molecules such as ammonia, is due to surface stabilization following surface reconstructions, which are analogous to the (7×7)-reconstruction of the bulk Si(111) surface and the (2×1)-reconstruction of the Si(100) surface. Such reconstructions use up some of the dangling silicon bonds in forming extra surface Si–Si bonds, thereby reducing the tendency to adsorb and react with small molecules. The surfaces of these clusters have both five- and six-membered rings. Although Kaxiras's models are based on four-coordinate interior atoms, DFT calculations by Röthlisberger, Andreoni and Parrinello indicate that clusters of around this size have dense cores composed of silicon atoms with coordination numbers greater than four.

6.3.6 Could Si Clusters be Metallic?

Many semiconducting elements have significantly reduced resistivity upon melting and some, such as silicon and germanium, actually become metallic. This metallic conductivity (which is essentially due to band overlap) is associated with an increase in density on melting and is consistent with the fact that these elements also become metallic at high pressures. The increase in density of these covalently bonded network solids on melting is due to the increase in mean coordination number and decrease in average network ring size which accompany the collapse of the open network.

Cluster polarizability measurements by Becker, Hensel and co-workers revealed that Si clusters in the range $25 \leq N \leq 50$ have size-dependent electric polarizabilities. These experiments involve the deflection of a beam of neutral clusters by an inhomogeneous magnetic field, where the deflection (d) is related to the static dipole polarizability (α) according to:

$$d = C \frac{\alpha E^2}{Mv^2} \tag{6.10}$$

where E is the electric field strength, M is the mass of the cluster and v is its velocity. C is a constant which depends on the experimental apparatus. The polarizability per atom ($\alpha_N = \alpha/N$) varies from 1.9 Å^3 for $N \approx 25$, to 3.2 Å^3 for $N \approx 35$ and back down to 1.9 Å^3 for $N \approx 50$–60. The lower polarizabilities ($\alpha_N = 1.9$ Å^3) correspond to dielectric constants ($\varepsilon = 3.2$) which are considerably smaller than those for bulk silicon ($\varepsilon_B = 11.8$). By contrast, the higher polarizability ($\alpha_N = 3.2$ Å^3 at $N \approx 35$) corresponds to a dielectric constant which is comparable with that of bulk silicon.

Interestingly, this peak in polarizability at $N \approx 35$ occurs in the size range where DFT calculations have predicted structures containing several highly coordinated core atoms, and a band gap of only a few tenths of an eV. As it is generally true that low band gaps lead to high polarizabilities, Becker and Hensel have interpreted their results in terms of a two-shell dielectric model, where clusters such as Si_{35} consist of an inner (core) sphere of highly coordinated atoms, with a high dielectric constant ($\varepsilon_C = 50$), surrounded by an outer shell of surface atoms, which has the same low dielectric constant ($\varepsilon_S = 50$) as that found for larger clusters ($N \approx 60$).

There is therefore a possibility of clusters which have insulating exteriors (crusts) and metallic cores, as shown schematically in Figure 6.14. This coexistence of metallic and nonmetallic regions is analogous to the coexistence of liquid-like and solid-like regions, which has been postulated by Berry, on the basis of Molecular Dynamics simulations, for noble gas and metal clusters. This coexistence, which is a finite size effect, generally results in solid-like (rigid) cores and liquid-like (mobile) exteriors. Berry has suggested that elements which are denser in the liquid phase than in the solid (such as Si and Ge) may give rise to clusters in which this situation is reversed—i.e. where the core is molten and the outer regions of the cluster form a solid crust. Since molten Si and Ge are metallic, it is interesting to speculate on whether in the experiment of Becker and Hensel the cores of the clusters are metallic because they are solid-like but dense or because they are actually molten. The answer will depend on the experimental conditions under which the clusters are generated and studied and, in particular, the cluster temperature.

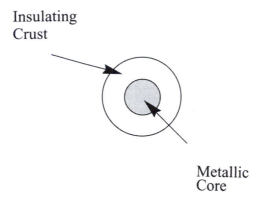

Insulating
Crust

Metallic
Core

Figure 6.14 Schematic representation of a silicon cluster with a high polarizability, due to having a dense, metallic core and a less dense, insulating mantle.

6.3.7 Photoelectron Spectroscopy

Photoelectron spectroscopy measurements have been performed on anionic silicon clusters. The clusters Si_4^-, Si_7^- and Si_{10}^- exhibit low-electron-binding energy peaks, indicating that the corresponding neutral clusters are closed shell species, with HOMO–LUMO gaps in the range 1–1.5 eV. (The corresponding band gap in bulk silicon is 1.09 eV.) The high HOMO–LUMO gap for Si_{10} is consistent with the relatively high ionization energy of this cluster, and with its high abundance in photofragmentation studies.

6.3.8 Optical Properties of Silicon Clusters

The evolution of the electronic states in silicon is readily observed by determining the size dependence of the optical properties of silicon clusters. Although ion mobility and chemical reactivity measurements indicate that there are at least two isomers for silicon clusters with around 30 atoms, the optical spectra of Si_{18}–Si_{41} are found to vary little with nuclearity. Some spectral features are found which are also present for bulk silicon. Photoluminescence in silicon (and other semiconductor) clusters will be discussed in Section 8.3.

6.3.9 Hydrogenated Si Clusters

Decomposition of the gaseous silanes (the silicon equivalent of the alkanes) Si_NH_{2N+2}, either over a heated wire filament or by radio frequency discharge leads to the formation of hydrogenated silicon clusters, as well as hydrogen-free clusters:

$$SiH_4 \text{ (1410°C/W filament)} \rightarrow Si_N + Si_NH_x + \dots$$

Analysis of the mass spectra of the clusters generated in this way show that hydrogen rich ($Si_NH_x^+$, $x > N$) clusters predominate for $N \lesssim 11$. There is an even–odd alternation, with clusters having an odd number of hydrogen atoms giving rise to larger peaks—probably because the corresponding cationic clusters $Si_NH_x^+$ have an even number of electrons. As the clusters get larger the ratio x/N decreases—peaking at around $x/N = 1$ (i.e. for $(SiH)_N^+$—although a range of x values is still found for each value of N. The $Si_{25}H_x^+$ series of clusters can have x = 5–30 hydrogen atoms.

Theoretical calculations have shown that the hydrogen atoms in these clusters saturate the silicon cluster by taking up the dangling surface bonds. This lowers the cluster surface energy and prevents surface reconstruction (which normally occurs so as to lower the energies of the dangling bond electrons). The presence of the hydrogen atoms, as the clusters grow, favours the retention of pseudo-tetrahedral bonding arrangements around the silicon atoms (i.e. the sp^3 hybridization of the silicon atoms in the silanes is retained in the clusters). This leads to clusters with lower silicon atom coordination numbers and larger (five- and six-membered) silicon rings (and therefore less condensed structures) than for the naked silicon clusters.

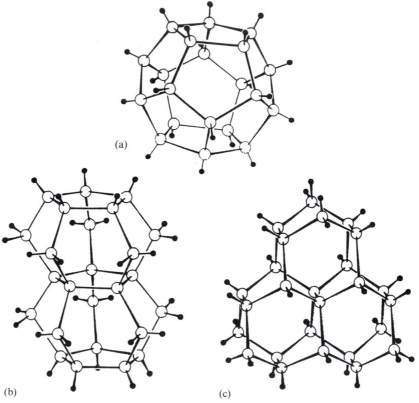

Figure 6.15 Structures proposed for neutral hydrogenated silicon clusters: (a) $Si_{20}H_{20}$; (b); $Si_{25}H_{30}$; (c) $Si_{26}H_{30}$.

Some structures which have been postulated for hydrogenated silicon clusters are shown in Figure 6.15. The structure proposed for $Si_{20}H_{20}$ (Figure 6.15a) is a dodecahedron with pentagonal faces and Si–Si–Si bond angles which are close to the ideal tetrahedral bond angle for sp^3 hybridized Si. (The analogous hydrocarbon molecule, dodecahedrane, was synthesized by Pacquette and co-workers as the result of a long, but elegant organic chemistry synthesis.) The postulated structure for $Si_{25}H_{30}$ (Figure 6.15b) possess both five- and six-membered rings and that for $Si_{26}H_{30}$ (Figure 6.15c) consists of six-membered rings in both the chair and the boat conformations. The proposed structure of $Si_{26}H_{30}$ resembles a fragment of the diamond-like bulk silicon structure—though the diamond structure consists of only chair shaped six-membered rings. Thus, for hydrogenated clusters, bulk-like structures are observed even at relatively low nuclearities—in contrast to the denser bare silicon clusters.

6.4 OTHER ELEMENTAL SEMICONDUCTOR CLUSTERS

6.4.1 Germanium Clusters

Mass spectral studies of germanium clusters give very similar results to silicon, with relatively high abundances observed for Ge_N^+, with $N = 6, 10, 14, 15$ and 18. The results of photofragmentation studies are again similar to those of silicon clusters, though there are a number of differences. Low intensity (0.5 mJ cm^{-2}) photofragmentation of Ge_{30}^+, for example, yields Ge_{20}^+ and Ge_{10}^+ (with this smaller cluster believed to arise from secondary fragmentation of Ge_{20}^+) as the only charged fragments, while high intensity (17.5 mJ cm^{-2}) irradiation leads to explosion into many smaller fragments. Structures have been proposed for these relatively stable ten- and twenty-atom clusters, invoking bond angles of approximately 109° and diamond-like geometries, though these structures are in disagreement with the results of density functional theory calculations.

Cluster ion mobility studies by Jarrold and co-workers indicate that, as for silicon, Ge_N^+ clusters adopt prolate geometries up to around $N = 35$, beyond which size more spherical oblate geometries predominate. The prolate structures have again been attributed to stacks of tricapped trigonal prisms.

As for silicon, the cluster Ge_{10} has a high ionization energy, by comparison with its neighbours, which is consistent with the photodissociation measurements. The photoelectron spectra of anionic germanium clusters are very similar to those of silicon, suggesting that the bonding in these closely related species is rather similar.

6.4.2 Tin and Lead Clusters

On descending a group in the *p*-block region of the periodic table, the metallic character of the elements is found to increase. For group 14 the semiconducting band gap decreases on going from carbon to silicon and germanium. While the heaviest member of the group, lead, is clearly a metal, the situation for tin is more complicated. The most stable allotrope of tin under ambient conditions is white tin (the β-phase), which is metallic. However at low temperatures tin undergoes a phase transition to the grey (α-Sn) allotrope, which is a narrow-gap semiconductor. Tin, therefore, can be said to lie at the border between metallic and non-metallic (semiconducting) elements. For this reason, it is interesting to see how tin clusters compare, as regards their structures and properties, with those of silicon, germanium and lead.

The mass spectrum of tin clusters possesses fewer features (intensity fluctuations) than observed for silicon and germanium, though the intensity of the peak corresponding to Sn_{15}^{+} is much larger than that for Sn_{14}^{+}. For lead, there are magic numbers corresponding to Pb_{N}^{+} with $N = 7$, 10, 13 and 19 and a deep minimum at $N = 14$. As discussed in Chapter 2, the peaks at $N = 7$, 13 and 19 are characteristic of icosahedral growth. Icosahedral cluster growth is commonly found for simple metal and transition metal clusters (such as Ca_N and Co_N) in the geometric shell size regime (see Chapter 4) and, as such is consistent with the metallic nature of elemental lead.

Recently, Shvartsburg and Jarrold have compared the mobilities of silicon, germanium, tin and lead clusters, from drift tube experiments. Figure 6.16 shows the relative mobilities of cluster cations (E_{N}^{+}, with $N \leq 32$ and $E = Si$, Ge, Sn and Pb). The relative mobility $K_{rel} = K_{exp}/K_{sph}$, where K_{exp} is the experimental mobility of cluster E_{N}^{+} and K_{sph} is the mobility calculated for a hypothetical sphere of volume $V_{sph} = NV_{at}$ (V_{at} is the atomic volume of element E under ambient conditions). For Si, Ge and Sn the K_{rel} values are reasonably similar up to $N = 25$ and are consistent with prolate cluster geometries, as discussed in Section 6.3.3. At $N = 25$ the more spherical (and more mobile) oblate isomers are observed for Si, in addition to the prolate isomers. Under the conditions of these experiments, no spheroidal Ge or Sn clusters are observed up to $N = 32$.

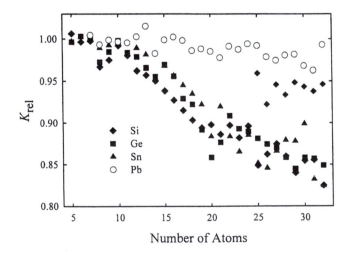

Figure 6.16 Relative mobilities of group 14 clusters. Filled symbols: Si (diamonds); Ge (squares); Sn (triangles). Empty circles: Pb.

Figure 6.16 shows very clearly that lead clusters have high relative mobilities ($K_{rel} \approx 1$), even for low nuclearities, and show little variation with cluster size (the slight overall decrease in K_{rel} as a function of n is due to multiple scattering). This leads to the conclusion that the Pb clusters have pseudo-spherical topologies throughout this size range. Particularly noticeable is the very high mobility of Pb_{13}^+ and the low mobility of Pb_{13}^+. This is consistent with a centred icosahedral structure for Pb_{13}^+, which is a very compact spherical shape, and also agrees with the mass spectrum, where a large Pb_{13}^+ peak was measured. It is also apparent that, for $N \geq 25$, the relative mobilities of the Pb clusters are greater than those of the pseudo-spherical Si clusters. The lower K_{rel} values of the Si clusters (≈ 0.95) indicate a less dense packing, consistent with covalent bonding and a network structure, rather than a pseudo-close-packed metallic structure.

These mobility experiments show a similarity between the mobilities of silicon, germanium and tin clusters (though the slightly higher mobilities of tin clusters may indicate slightly denser packing) and a clear distinction between these elements and lead. Thus, for the clusters of group 14 elements, the transition from covalent to metallic bonding occurs between tin and lead, rather than between germanium and tin. Bearing in mind the change of cluster bonding with number of atoms (see Chapter 5), this situation is likely to change as the cluster nuclearity increases. Indeed, nanometre-sized Sn particles are found to exhibit bulk-like metallic packing.

Interestingly, by studying the temperature-dependence of cluster mobility, Shvartsburg and Jarrold have shown that tin clusters (with 10–30 atoms) remain solid until approximately 50 K above the buk melting temperature. This unusual phenomenon (see Chapter 1 for a discussion of size-dependent trends in cluster melting) is possibly linked to the fact that, in this size regime, the tin clusters adopt structures which are quite distinct from that of bulk β-Sn.

6.4.3 Boron Clusters

In the solid state, many boron compounds consist of three-dimensional networks of boron clusters. Metal borides MB_6 have arrays of B_6 octahedra, while compounds with the formula MB_{12} generally consist of B_{12} cuboctahedra. Solid elemental boron also consists of 3-D networks of linked icosahedral B_{12} clusters and a building block of the structure of rhombohedral β-boron is a B_{84} unit, consisting of a C_{60}-like B_{60} cage, which is bonded to an encapsulated B_{12} icosahedron via twelve linking B atoms. Since the pioneering work of Alfred Stock in the 1930s, a rich chemistry has been developed of boron hydride clusters, where each boron atom is bonded to at least one hydrogen ligand. Many of these clusters are based on the icosahedral cluster $[B_{12}H_{12}]^{2-}$ and many derivatives are known with main group (e.g. carbon) and transition metal atoms incorporated into the boron cluster cage.

Despite the rich chemistry of ligated, inorganic boron clusters, there have been relatively few studies of isolated bare boron clusters. In 1964, Chupka and Berkowitz observed the formation of small boron clusters after vaporization of solid boron using a ruby laser. Anderson and co-workers have performed a number of studies of cationic boron clusters generated by laser vaporization. In these experiments, the cluster ions are trapped and cooled in an octopole ion guide and size-selected using a Wien filter before injecting the ions into a second octupole ion guide. Using this cluster source, clusters with 5, 10, 11 and 13 atoms are found to have high MS abundances.

Collision-induced cluster fragmentation studies have shown that B_5^+ and B_{13}^+ are particulalry stable (i.e. they have high bond dissociation energies), while B_{10}^+ and B_{11}^+ are no more stable than B_{12}^+, which has low relative abundance in the mass spectrum. B_5^+ is also found as an abundant fragment arising from collision induced dissociation. Calulations indicate that the boron clusters are generally close-packed with B_5^+ having a trigonal bipyramidal geometry and it has been conjectured that B_{13}^+ has either a centred icosahedral or a centred cuboctahedral geometry.

Anderson and co-workers have also studied the reaction of boron clusters with small molecules, such as O_2, D_2 and D_2O. Reaction with O_2 results in cluster fragmentation and the generation of both B_N^+ and B_NO^+ products.

It has been shown (see Chapter 1) that icosahedral clusters are destabilized at high nuclearities because there is a bulk strain associated with the (Mackay) icosahedron—i.e. it is impossible to construct an icosahedron from regular tetrahedra without introducing gaps into the structure. Interestingly, in 1998 O'Keeffe and colleagues showed that boron suboxide clusters, with the stoichiometry B_6O, form Mackay icosahedral structures with up to 10^{14} atoms— with cluster diameters of nearly 10 μm! These microscopic icosahedra (an example of which is shown in Figure 6.17, along with the corresponding electron diffraction pattern) are stabilized because the boron suboxide clusters are built up from irregular tetrahedra, which match the angle at the centre of the icosahedron ($4\pi/20$ radians) exactly.

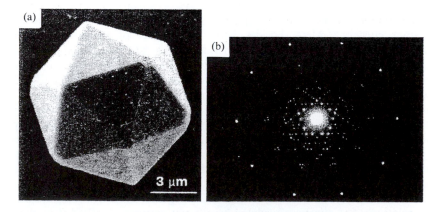

Figure 6.17 (a) Electron micrograph of a 10 μm diameter $(B_6O)_N$ Mackay icosahedron. (b) Electron diffraction pattern of this cluster, showing perfect five-fold symmetry.

6.5 COMPOUND SEMICONDUCTOR CLUSTERS

6.5.1 Introduction

Many semiconducting solids are compounds rather than elements. Examples of these are the so called III–V compounds formed between the elements of groups 13 and 15, of which GaAs is probably the most widely used in the electronics industry. Another important class of semiconductors are the II–VI compounds formed between the elements of groups 12 and 16, with a common example being CdSe (see Chapter 7). The III–V and II–VI compounds have a 1:1 stoichiometry and are formally isoelectronic with the elements of group 14, since one component lies one (or two) places to the left of group 14 and the other lies one (or two) places to the right. Each movement to the left (right) leads to a change of −1 (+1) in the number of valence electrons. The III–V and II–VI compound semiconductors generally have structures based on those of the elemental semiconductors, but with an alternating arrangement of the two types of atoms. Thus, the most stable form of GaAs has the cubic diamond structure (the same as its isoelectronic counterpart germanium), with an alternating Ga–As arrangement. This structure is known as the Zinc Blende structure—after one of the allotropes of ZnS (a II-VI compound). One of the allotropes of BN is composed of hexagonal sheets, with alternating B and N atoms, as in the graphite form of the isoelectronic element carbon.

The alternating A–B pattern generally observed for compound semiconductors AB is preferred due to the polar nature of these compounds. Thus, although the bonding in a compound such as GaAs is primarily covalent, the electronegativity difference between the elements (As > Ga) means that the Ga–As bond is polarized as: $Ga^{\delta+}$–$As^{\delta-}$. It should be noted that, although the formal oxidation states of the elements in GaAs are Ga^{3+} and As^{3-}, these compounds are closer to being covalent than ionic. However, as the difference in electronegativity increases (e.g. for metal halides and oxides), the bonding will become more ionic. Ionic clusters will be discussed in Chapter 7.

The relationship between the electron counts and structures of bulk elemental and compound semiconductors also holds for their clusters. Thus, the gallium arsenide cluster $(GaAs)_N$ is isolelectronic with the germanium cluster Ge_{2N} and $(BN)_N$ is isoelectronic with C_{2N}. The polarity of the bonding in these heteronuclear clusters, however, may cause differences in structures, so as to avoid having two atoms of the same type adjacent to each other. (This would rule out fullerene-like structures for $(BN)_N$ clusters, for example.) In this section, I will deal briefly with a few examples of compound semiconductor clusters.

6.5.2 Gallium Arsenide Clusters

On the basis of microwave spectroscopic measurements, Morse and co-workers have concluded that the bond length of the GaAs molecule is greater than that in bulk GaAs. This unusual finding—for example the bond lengths in the molecules

BN, C_2, Si_2, Sn_2 and Pb_2 are all shorter than in the bulk—has been attributed to the greater ionic contribution to the bonding in solid GaAs compared with the GaAs molecule.

Smalley and his co-workers have investigated the photodissociation of cationic $[Ga_xAs_y]^+$ and anionic $[Ga_xAs_y]^-$ clusters, in an analogous fashion to their work on silicon clusters. They found that, in contrast to silicon, GaAs clusters tend to dissociate by loss of individual atoms (as for simple metal clusters), rather than by losing 'magic number' fragments. The polar nature of the bonding in GaAs clusters is reflected in the fact that the reaction with HCl results in etching, via removal of gallium atoms as GaCl.

$$[Ga_xAs_y]^- + HCl \rightarrow [Ga_{x-1}As_yH]^+ + GaCl$$

Theoretical calculations on Ga_xAs_y clusters predict structures which are generally similar to those of silicon or germanium. Unlike ionic clusters, where it is very unfavourable to have ions of the same sign charge adjacent to one another, the less-polar (more covalent) GaAs clusters can have Ga–Ga and/or As–As bonds.

6.5.3 Indium Phosphide Clusters

Photodissociation measurements on the III–V clusters In_5P_y–In_9P_y (with $y = 1$–5) have revealed dissociation energies in the range 2.0–2.4 eV. By measuring the two-photon dissociation spectra of these clusters, it is possible to map out their IR spectra. It was found that the stoichiometric (1:1) InP clusters In_3P_3–In_6P_6 show discrete electronic absorptions with energies much smaller than the bulk band gap (1.18 eV), while continuous absorption is observed with a threshold in the range 1.3–1.5 eV—i.e. slightly higher than the bulk band gap. These results have been interpreted in terms of a bonding model similar to that in bulk InP, with a filled valence band of In–P bonding states, an empty valence band of In–P antibonding states and some states in the band gap, which are either In–In bonding or dangling bond states.

6.6 EXERCISES

6.1   List the factors which influence the stabilities of fullerene isomers.

6.2   Explain, with examples how ^{13}C NMR can be used to distinguish between alternative fullerene isomers for a given number of carbon atoms. (If you are feeling adventurous, try predicting the ^{13}C NMR spectra of the fullerene isomers in Figure 6.4.)

6.3   Describe how ion mobility experiments can be used to provide information on the structures and relative stabilities of silicon cluster isomers.

6.7 FURTHER READING

Attfield, J.P., Johnston, R.L., Kroto, H.W. and Prassides, K., 1999, New science from new materials. In *The Age of the Molecule*, edited by Hall, N., (London, Royal Society of Chemistry), pp. 181–208.

Bowers, M.T., 1994, Cluster ions: carbon, met-cars, and σ-bond activation. *Accounts of Chemical Research*, **27**, pp. 324–332.

Brus, L., 1994, Larger semiconductor clusters ('quantum dots'). In *Clusters of Atoms and Molecules*, Vol.II, edited by Haberland, H., (Berlin, Springer-Verlag), pp. 312–320.

Campbell, E.E.B., 1994, Carbon clusters. In *Clusters of Atoms and Molecules*, Vol.I, edited by Haberland, H., (Berlin, Springer-Verlag), pp. 331–356.

Fowler, P.W. and Manolopoulos, D.E., 1995, *An Atlas of Fullerenes*, (Oxford, Oxford University Press).

Jarrold, M.F., 1989, Metal and semiconductor cluster ions. In *Gas Phase Inorganic Chemistry*, edited by Russell, D.H., (New York, Plenum), pp. 137–192.

Kroto, H.W., 1988, Space, stars, C_{60} and soot. *Science*, **242**, pp. 1139–1145.

Kroto, H.W., Allaf, A.W. and Balm, S.P., 1991, C_{60}—Buckminsterfullerene. *Chemical Reviews*, **91**, pp. 1213–1235.

Kroto, H.W., Heath, J.R., O'Brien, S.C., Curl, R.F., Smalley, R.E., 1985, C_{60}—Buckminsterfullerene. *Nature*, **318**, pp. 162–163.

Kroto, H.W., Walton, D.R.M. (editors), 1993, *The Fullerenes: New Horizons for the Chemistry, Physics and Astrophysics of Carbon*, edited by (Cambridge, Cambridge University Press).

Martin, T.P., 1996, Shells of atoms. *Physics Reports*, **273**, pp. 199–241.

Phillips, J.C., 1986, Chemical bonding, kinetics, and the approach to equilibrium structures of simple metallic, molecular and network microclusters, *Chemical Reviews*, **86**, pp. 619–634.

Scuseria, G.E., 1996, *Ab initio* calculations of fullerenes. *Science*, **271**, pp. 942–945.

Shvartsburg, A.A., Hudgins, R.R., Dugourd, P. and Jarrold, M.F., 2001, Structural information from ion mobility measurements: applications to semiconductor clusters. *Chemical Society Reviews*, **30**, pp. 26–35.

Sugano, S., 1991, *Microcluster Physics*, (Berlin, Springer-Verlag), pp. 104–117.

CHAPTER SEVEN

Ionic Clusters

7.1 INTRODUCTION

In this chapter, the term 'ionic cluster' will be used to refer to clusters derived from ionic solids—i.e. compounds formed between elements which have a large difference in electronegativity, such that a charge-separated description (e.g. A^+B^-) of the bonding becomes appropriate. Such a description is particularly valid for the singly charged alkali metal halides, but, as we shall see, may not be so appropriate for oxide clusters, where anion polarization and covalency may play a role in the bonding. Ionic compounds form between electropositive metals and electronegative non-metals—examples being the halides and oxides of groups 1 and 2 (the alkali metals and alkaline earth elements)—and are widespread in inorganic solid state chemistry. They have been widely studied and serve as model materials, due to their simple electrostatic bonding and the ease of sample preparation.

As the difference in electronegativity between the positive and negative elements decreases, the distinction between ionic compound and semiconductor becomes blurred. Thus, chalcogenide compounds such as CdSe may be regarded as an ionic compound, with a significant amount of covalency in the bonding (and a narrow insulating band gap), or as a semiconductor cluster in which the Cd–Se bonds are polarized (i.e. the electrons in the bonds are not shared equally). A number of compound semiconductors, such as GaAs, have been discussed in Chapter 6.

Ionic clusters may be generated with positive or negative charge, or they may be neutral (uncharged). Ionic clusters can be generated by quenching the vapour, resulting from heating or laser vaporization of ionic compounds, in a stream of a cold inert gas. The study of ionic clusters allows the possibility of determining the size at which ionic clusters begin to take on the properties of the bulk (i.e. when they may be described as 'nanocrystalline') and how the interionic bonding changes as a function of size. As well as being ideal model systems for studying size-dependent properties, there is much interest in ionic nanoclusters in their own right. Thus, the silver halides are of intrinsic interest because of their relevance to the photographic process, while nanoclusters of the group 12 chalcogenides have a number of applications—for example as gas sensors (ZnS) and photodetectors (CdS, CdSe).

Finally ionic clusters are even of environmental relevance. The simple sodium halide clusters NaCl and NaBr are important in the marine atmosphere—forming from salt-water droplets which are swept up from the sea and dehydrated by the Sun's action. The reaction of these atmospheric 'microbrine' clusters with nitrogen oxide (NO_x) pollutants can give rise to chlorine and bromine atoms, which in turn are responsible for catalytic ozone depletion.

7.2 BONDING IN IONIC CLUSTERS

7.2.1 Rigid Ion Model

The simplest bonding model which can be applied to ionic solids (and their clusters) is the *rigid ion model*, in which the ions carry fixed charges and have fixed sizes. In the rigid ion model, the total interaction energy for a cluster consisting of N ions is given by the sum of the long-range electrostatic Coulomb potential (which may be attractive or repulsive, depending on the signs of the ionic charges) and a short range repulsive Born–Mayer potential, which reflects the short range repulsive energy due to overlap of the electron density of the ions:

$$V(N) = \sum_{i,j}^{N} \left[\left(\frac{e^2}{4\pi\varepsilon_0} \right) \left(\frac{q_i q_j}{r_{ij}} \right) + A_{ij} \exp\left(-r_{ij} / \rho_{ij} \right) \right] \tag{7.1}$$

where q_i and q_j are the charges on ions i and j, r_{ij} is the inter-ionic separation and the Born–Mayer parameters (A_{ij} and ρ_{ij}) also depend on the nature of ions i and j.

The long range ($1/r$) nature of the Coulomb electrostatic potential and the strength of the attractive and repulsive interactions results in highly ordered ionic clusters, even at low nuclearities—in contrast to metal and semiconductor clusters. Due to the strong repulsive interactions, ions with the same sign charge are separated from each other. Thus, ionic clusters, are composed of even-membered rings, with alternating anions and cations.

7.2.2 Polarizable Ion Model

As ions (especially large anions, such as I^- and O^{2-}) are not rigid, but can be deformed (polarized) due to the presence of the other ions in the cluster (in particular the cations), Equation 7.1 has to be modified to include higher-order electrostatic terms. A general expression can be written for the total potential energy of an N-ion cluster:

$$V(N) = \sum_{i}^{N} \left[\frac{\mu_i^2}{(4\pi\varepsilon_0)\alpha_i} + \frac{1}{2} \sum_{j\neq i}^{N} \left(V_{ij}^R + V_{ij}^{MD} + V_{ij}^{DD} \right) \right] \tag{7.2}$$

where μ_i is the dipole moment on ion i and α_i is its polarizability. The first term in Equation 7.2 represents the self-energy of the ionic dipole. The term V_{ij}^R is the rigid ion potential (as in Equation (7.1)), but with the distance r_{ij} in the Born-Mayer repulsive potential replaced by an effective distance (r_{ij}^{eff}) to take into account the deformation of the electronic shells of the ions upon coordination. The final terms in Equation (7.2) are monopole–dipole (V_{ij}^{MD}) and dipole–dipole (V_{ij}^{DD}) interaction energies, which are defined as:

$$V_{ij}^{MD} = -\left(\frac{e}{4\pi\varepsilon_0} \right) \left(\frac{q_i \mu_j + q_j \mu_i}{r_{ij}^2} \right) \tag{7.3}$$

and

$$V_{ij}^{DD} = \left(\frac{1}{4\pi\varepsilon_0}\right)\left(\frac{4\mu_i\mu_j}{r_{ij}^3}\right) \tag{7.4}$$

More sophisticated extensions of the ionic model include explicit coordination number and density-dependent terms and the inclusion of dispersion forces, which can be quite large, especially for large anions such as iodide.

7.3 ALKALI HALIDE CLUSTERS

7.3.1 Structures

Mass spectroscopic studies by Dunlap and co-workers at the Naval Research Laboratory in 1981 showed that the most abundant (and presumably the most stable) alkali halide nanoclusters are those with exactly the right number of ions to fill the (bulk-like) cubic sites in a structure with cuboidal morphology. Thus, the stable clusters correspond to cuboidal fragments of the bulk rock-salt (NaCl) structure, in which each ion is octahedrally coordinated to six ions of opposite charge, so that there is an alternating arrangement of cations and anions, as shown in Figure 7.1.

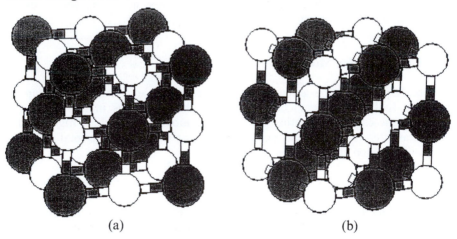

(a)   (b)

Figure 7.1 (a) The (3,3,3) ionic cluster $[Na_{14}Cl_{13}]^+$ (dark spheres = Na^+, light spheres = Cl^-). (b) The (3,3,3) ionic cluster $[Na_{13}Cl_{14}]^-$.

These cuboidal nanocrystals are stable because they only expose square (100) surfaces (this being the lowest energy surface of the rock-salt structure). Such cuboidal magic numbers occur when the total number of ions (n) is the product of three integers (the number of ions along each edge): $m = i{\times}j{\times}k$. For even values of n, this condition may be satisfied by neutral stoichiometric (1:1) clusters $(MX)_N$ (i.e. with $N = m/2$). When m is odd (i.e. when all of (i,j,k) are odd) then there must be one more of one type of ion than the other—giving rise to charged clusters $[M_{N+1}X_N]^+$ or $[M_N X_{N+1}]^-$. In contrast to metal or semiconductor clusters, the charge in non-stoichiometric ionic clusters is localized on the ions, since, for example, $[M_{N+1}X_N]^+$ can be rewritten as $[(M^+)_{N+1}(X^-)_N]^+$. Examples of this type of cluster are the 27-ion (3,3,3) clusters $[Na_{14}Cl_{13}]^+$ (Figure 7.1a) and $[Na_{13}Cl_{14}]^-$ (Figure 7.1b).

Jarrold and co-workers have performed ion mobility studies (see Chapter 1) of alkali halide clusters. For the 71-ion cluster $(NaCl)_{35}Cl^-$, for example, they have distinguished three isomers, with differing drift times, as shown in Figure 7.2. The 'fastest' cluster has been identified as a (5,5,3) cuboid, with four ions missing. This isomer is more nearly spherical than the 'slower' (6,4,3)–1 and (8,3,3)–1 defective cuboids, where the negative numbers indicate the number of missing ions. (Note that, although these structures are described as defective cuboids, they are still comprised of Na^+ and Cl^- ions and should not be confused with the excess electron clusters described in Section 7.3.4.)

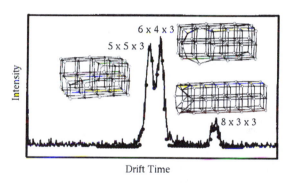

Figure 7.2 The results of ion mobility measurements on $(NaCl)_{35}Cl^-$. Proposed structures are included next to the relevant peaks. These are actually defect structures and the *i×j×k* labels refer to the parent complete cuboidal clusters.

Calculations by Doye and Wales on $(NaCl)_{35}Cl^-$, using an empirical ionic potential, show that there are families of isomers, often closely spaced in energy, based on these defective cuboids, as well as a family of (5,4,4)–9 isomers. Although the isomer energies depend on the positions of the defects, Doye and Wales's calculations predict the best '5×5×3' structures to be lower in energy than the '6×4×3' and '8×3×3' geometries, with the highly defective '5×4×4' structures being least stable (which is consistent with their non-appearance in Jarrold's ion mobility experiments). Jarrold has observed time dependent changes in peak intensities, which reflect time dependent changes in the populations of the various structure types. These observations show that the '5×5×3' structures are indeed lowest in energy, consistent with the calculations of Wales. Finally, Doye and Wales have shown that these NaCl cluster isomers can be interconverted by highly cooperative mechanisms, wherein planes of atoms glide past each other, rather than via a surface process (as postulated by Jarrold). Interestingly, this means that NaCl clusters (or 'nanocrystals') are plastic, not brittle like the bulk—another example of a finite size effect.

Figure 7.3 shows the mass spectrum of $(NaI)_N Na^+$ clusters, due to Martin and co-workers. As these clusters consist of $(2N+1)$ ions, magic numbers (peaks in the spectra), corresponding to perfect $i \times j \times k$ cuboids, are only possible when the numbers of atoms along each edge (i,j,k) are all odd numbers. Cuboidal clusters with $i \times j \times k$ = even are therefore absent. The marked features are due to cubic (ℓ,ℓ,ℓ) clusters, with ℓ ions along each edge, where $\ell^3 = 2N+1$: $N = 62$ corresponding to a (5,5,5) cube and $N = 364$ to a (9,9,9) cube, for example. Other intense peaks are due to non-cubic, cuboidal (i,j,k) geometries $(i \times j \times k$ = odd). Jarrold's experiments have shown, however, that there may actually be more than one isomer present for a particular nuclearity.

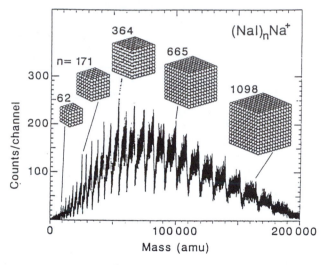

Figure 7.3 Mass spectrum of $(NaI)_N Na^+$ clusters. Features due to complete cubes are indicated. Other significant features correspond to cuboidal clusters.

Martin and colleagues have calculated the lowest energy structures of neutral stoichiometric clusters $(MX)_N$ (MX = NaCl, NaI, CsF and CsI), for $N < 20$ ion pairs. In this size range, Martin found a competition between cubic and cuboidal structures based on stacking $M_2 X_2$ squares and structures based on stacked $M_3 X_3$ hexagons, as shown in Figure 7.4. Finite size effects are manifest, as $(NaCl)_N$ clusters are predicted to be hexagonal stacks for $N = 6, 9, 12$ and 15, rather than adopting cubic or cuboidal structures based on the bulk rocksalt structure. For $N = 18$, however, a bulk-like structure is found. Hexagonal building blocks are also found in many $(NaI)_N$ clusters, though for $N = 12$, the truncated octahedron is lower in energy than a stack of four hexagons. In the cases of both $(CsF)_N$ and $(CsI)_N$ clusters, cubic rock-salt structures are preferred over hexagonal structures in this size regime. This also shows the importance of finite size effects, since in the bulk phase, CsF and CsI have different structures: CsF has the rock-salt NaCl structure, while CsI has the CsCl structure (which may be regarded as being derived from body-centred cubic packing), where each ion (Cs^+/I^-) lies at the centre of a cube of the other type of ion.

Martin and co-workers have also performed calculations on small alkali halide clusters with an excess metal cation, resulting in a positively charged cluster $(MX)_N M^+$. A variety of structures were observed, depending on whether rigid or polarizable ion potentials were employed. One noticeably stable cluster, found for both models is the (3,3,3) cube for $N = 13$ ($[M_{14}X_{13}]^+$), which has been mentioned above as a significant feature in the mass spectra of alkali halide clusters. For NaCl clusters, other relatively large mass spectral peaks are found corresponding to $[Na_7Cl_6]^+$ and $[Na_{10}Cl_9]^+$. Martin's calculations indicate that these clusters also have low energy geometries containing Na_4Cl_4 cubic building blocks.

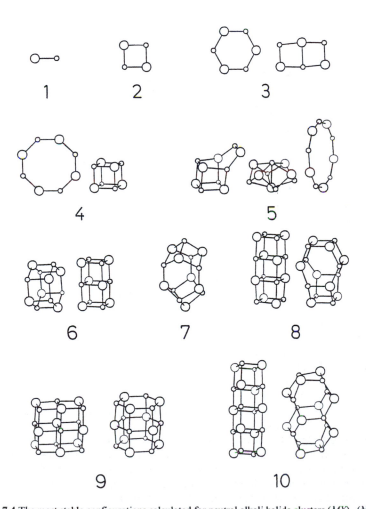

Figure 7.4 The most stable configurations calculated for neutral alkali halide clusters $(MX)_N$ ($N \le 10$).

7.3.2 Melting, Evaporation and Cleavage of Alkali Halide Clusters

One consequence of the stability of cuboidal ionic nanocrystals is that newly formed hot clusters tend to cool by evaporation of *MX* molecules, until a stable cuboidal nanocrystal is generated.

$$(MX)_N \rightarrow (MX)_{N-1} + MX \rightarrow (MX)_{N-2} + 2MX \rightarrow \dots$$

This process is analogous to the shattering of brittle crystals to generate facetted fragments and occurs because of impact-induced shear stresses.

Large clusters with $m \neq i{\times}j{\times}k$ tend to adopt nearly-cuboidal structures with the extra atoms forming terraces, analogous to steps on the surface of a crystal. When terrace structures are not possible, then calculations (using model ionic potentials) predict that surface reconstructions take place, giving rise to hexagonal, as well as square rings on the surface of the cluster.

Experiments in which charged cuboidal particles are accelerated onto a solid surface have shown that the clusters cleave into two clusters which are also perfectly cuboidal:

$$(MX)_{m/2} \ (m = i{\times}j{\times}k) \rightarrow (MX)_{p/2} \ (p = i'{\times}j'{\times}k') + (MX)_{q/2} \ (q = i''{\times}j''{\times}k'')$$

Large cuboidal clusters, such as $(3,5,5)$-$[Na_{38}F_{37}]^+$ can cleave in more than one place, so a variety of cuboidal fragments can be formed.

Clusters with terraces readily cleave off the terraces on impact, thereby generating a cuboidal cluster. On the other hand, impacting smaller clusters on solid surfaces (or heating them) leads to the loss of *MX* molecules and small clusters. Halide clusters can be molten by heating or applying a severe shock and the melting process is generally accompanied by evaporation of *MX* molecules.

7.3.3 Reactivities of Alkali Halide Clusters

Although most alkali halide clusters readily adsorb polar molecules, such as NH_3 and H_2O, Whetten has shown that the perfect cuboidal clusters—such as $(3,3,3)$ and $(3,3,5)$—do not adsorb ammonia under ambient conditions. The most reactive halide clusters are those (e.g. $[Na_{13}F_{12}]^+$ and $[Na_{22}F_{21}]^+$) which are related to perfect cuboids by the loss of a single MX unit. In such cases, the polar molecule effectively substitutes for the missing dimer unit. At lower temperatures, all alkali halide clusters (including perfect cuboids) adsorb polar molecules—with little size variation—as the reaction rate is now collision limited.

7.3.4 Excess Electrons in Defect Alkali Halide Clusters

The ionic nature of the bonding in alkali halide clusters led Landman (in 1985) to suggest that structures could be stabilized in which an excess electron substitutes for a missing halide ion in a defect cluster. It has subsequently been shown (by Whetten and co-workers) that such clusters can be generated by UV irradiation of alkali halide clusters, via loss of a neutral halogen atom:

$$[M_aX_b]^{(a-b)+} + h\nu \rightarrow [M_aX_{b-1}]^{(a-b)+} + X^0$$

or by cluster condensation in an excess of metal vapour, at low temperatures. Examples of alkali halide clusters with excess electrons are the clusters $[Na_{14}F_{12}]^+$ (which can alternatively be written as $(Na^+)_{14}(F^-)_{12}(e^-)$) and $[Na_{13}F_{13}]^-$ (which may be written $(Na^+)_{13}(F^-)_{13}(e^-)$). Both of these clusters adopt the (3,3,3) cuboidal structure, with the excess electron occupying a surface fluoride site. In general, three possible types of excess electron cluster are possible: cationic $[M_NX_{N-2}]^+$; neutral $[M_NX_{N-1}]$; and anionic $[M_NX_N]^-$. The excess electron is stabilized by electrostatic interactions with the neighbouring alkali metal cations and may be described as a solvated electron, by analogy with electrons solvated by polar molecules in, for example, alkali metal–ammonia solutions.

Experiments (by Whetten and co-workers) have shown that very low energies (approximately 3 eV for neutral clusters of the form Na_NF_{N-1} = $(Na^+)_M(F^-)_{N-1}(e^-)$) are required to remove the excess electron, while analogous studies on the charged clusters $[Cs_NI_{N-2}]^+$ = $(Cs^+)_M(I^-)_{N-2}(e^-)$ indicate electron detachment energies of only around 5 eV, which is the lowest reported energy for creating a small doubly charged molecule ($[Cs_NI_{N-2}]^{2+}$).

Excess electron clusters have an additional low-energy decomposition pathway involving the loss of a neutral metal atom, which has been neutralized by thermal-induced migration of the excess electron. Such a neutral atom will no longer be tightly bound to the surrounding anions and thus can readily be ejected:

$$[M_aX_b(e^-)]^{(a-b-1)+} + \Delta \rightarrow [M_{a-1}X_{b-1}(M^+:e^-)]^{(a-b-1)+} \rightarrow [M_{a-1}X_{b-1}]^{(a-b-1)+} + M^0$$

An example of such a reaction is the (surface) collision-induced reaction:

$$[Na_{14}F_{12}]^+ \rightarrow [Na_{13}F_{12}]^+ + Na^0$$

7.3.5 Spectroscopy and Photochemistry of Alkali Halide Clusters

Normal (defect-free) alkali halide clusters are colourless since—as for the bulk alkali halides—the energy difference between the occupied halide atomic orbitals and the unoccupied metal atomic orbitals corresponds to absorption of light in the UV region of the spectrum. Due to the localized nature of the bonding in ionic compounds, the excited electron is transferred from an anion to one of the neighbouring metal cations—neutralizing both ions $M^+X^- \rightarrow M^0X^0$—and leading to an intense UV charge transfer absorption.

Absorption energies of alkali halide clusters generally decrease with increasing cluster size, lying between the high energy absorption of the *MX* molecule and the lower energy absorption of the bulk *MX* crystal. As an example, the CsI dimer has its charge transfer band maximum at 3.8 eV and the CsI crystal has its first absorption maximum at 5.7 eV. For the (3,3,3) cluster $[Cs_{14}I_{13}]^+$, there is an onset of absorption at 5.2 eV. The lowest energy absorptions correspond to charge transfer from the more exposed halide ions. In the (3,3,3) cluster cited above, the lowest energy charge transfer band therefore corresponds to the transfer of an electron from an edge I^- to a neighbouring vertex Cs^+. Among the alkali halide clusters, the perfect cuboidal clusters generally have the highest charge transfer absorption energies, reflecting the higher average coordination of the surface ions.

   As mentioned above, UV irradiation can also lead to loss of the neutral halogen atom created in the charge transfer process. The direct loss of the neutral metal atom is less common as the electron is not strongly localized on the metal cation (forming an excess electron state where it can be stabilized by interaction with a number of neighbouring cations) and the loss of M^0 is in fact a thermally activated process (see above).

   Although defect-free alkali halide clusters are colourless, excess electron alkali halide clusters are coloured. The colour arises from absorption of near-IR and red photons ($hv = 1.2$–2 eV), which causes a transition of the bound excess electron between its ground (atomic s-like) state and the first excited (p-like) state. Such 'colour centres' are also found in bulk alkali halide crystals, in which trapped electron defects (with quantized energy levels) are generated by UV irradiation.

7.4 METAL OXIDE CLUSTERS

7.4.1 Photoionization Studies of Cs–O Clusters

The oxide clusters of the heavy alkali metals (such as caesium) exhibit a rich variation in composition and, consequently, in properties. Martin and co-workers performed photoionization measurements on a range of Cs–O clusters, distinguishing four types of Cs_xO_y cluster according to composition:

(i)   metal rich clusters $Cs_N - Cs_2(Cs_2O)_N$
(ii)   doped oxide clusters $Cs(Cs_2O)_N$ (with one excess Cs atom)
(iii) oxide clusters $(Cs_2O)_N$ (with bulk 2:1 stoichiometry)
(iv)   oxygen-rich clusters Cs_NO_M (with $N/M < 2$).

   Martin found that the ionization energies of metal-rich clusters decrease with increasing oxygen content—a result which at first sight seems counter-intuitive. Indeed, doped oxide clusters, with a single excess Cs atom, have very low ionization energies (typically around 2.5 eV, compared with 4 eV for bulk Cs). These low values are due to the presence of donor states lying just below the conduction band, arising from the excess caesium atoms. These findings are consistent with experiments on thin films of Cs_2O, where the thermal activation barrier to conductivity is greatly reduced on Cs doping. Surface-oxidized Cs, which has a very low work function, is used in the construction of photocathodes which are active in the infrared region.

   Martin's experiments have also revealed an even–odd alternation in the ionization energies of oxygen rich clusters, Cs_NO_M, with high IPs for even N and low IPs for odd N. This can be explained by thinking of oxygen-rich clusters as being composed of caesium atoms (Cs^0) and ions (Cs^+), oxide (O^{2-}) and peroxide (O_2^{2-}) building blocks. Thus the 'even' oxygen-rich Cs_8O_5 cluster may be written as a mixture of three stoichiometric oxide units and one peroxide unit $(Cs_2O)_3(Cs_2O_2)$, with no excess Cs atoms. On the other hand, 'odd' oxygen-rich clusters, such as $Cs_9O_5 = Cs(Cs_2O)_3(Cs_2O_2)$ have excess Cs atoms, which give rise to low ionization energies, as described above for doped oxide clusters. Thus, the odd oxygen-rich clusters can be described as doped oxide–peroxide mixtures.

Finally, Martin noted that both Cs_9 and $Cs_{11}O$ have unusually low ionization energies (2.5–2.7 eV). Assuming that the oxygen is present as oxide, $Cs_{11}O$ can be written as $(Cs^0)_9(Cs_2O)$, which (like Cs_9) has nine electrons in metal-based orbitals. From the jellium model (see Chapter 4), an alkali metal cluster with nine electrons has one electron over a filled jellium shell—resulting in a low ionization energy. By contrast, the cluster $Cs_{10}O = (Cs^0)_8(Cs_2O)$ has a closed shell eight-electron configuration and, consequently, a relatively high ionization energy (approximately 3 eV).

7.4.2 Magnesium Oxide Clusters

There have been many experimental and theoretical studies of clusters of magnesium oxide (MgO), perhaps because it is the simplest ionic oxide species— crystallizing in the rock-salt (NaCl) structure. It is hoped that a detailed understanding of the growth of $(MgO)_N$ clusters will lead to an improved understanding of epitaxial growth of MgO surfaces. MgO clusters have also been postulated as nucleation sites for particle formation in oxygen-rich regions of space.

In 1988–89, Saunders published mass spectra and collision-induced fragmentation data for sputtered MgO cluster ions, and found magic number peaks for stoichiometric $(MgO)_N^+$ clusters (since charged clusters are detected by the mass spectrometer) with $N = 6, 9, 12$ and 15. These results were interpreted in terms of clusters built from $(MgO)_3$ sub-units. In the early 1990s, Ziemann and Castleman studied MgO clusters using laser-ionization time-of-flight mass spectrometry. They found magic number clusters for $(MgO)_N^+$ at $N = 2, 4, 6, 9, 12, 15, 18, 24, 27$ and 30. An interval, corresponding to three MgO molecules, was again observed. Ziemann and Castleman's experiments indicate that rocksalt-like cluster geometries predominate above $N = 30$.

By varying the experimental conditions (e.g. by varying the flow rate of the carrier gas in a gas aggregation cluster source), Ziemann and Castleman also detected non-stoichiometric, Mg-rich $(MgO)_NMg^+$ and $(MgO)_NMg^{2+}$ cluster ions. In the case of the $(MgO)_NMg^{2+}$ clusters, enhanced stabilities (magic numbers) were observed for $N = 8, 11, 13, 16, 19, 22, 25$ and 27. Again the larger peaks are separated by three MgO units.

Many theoretical studies have been carried out on neutral and charged MgO clusters, using a variety of theoretical approaches. It is well known that the oxide anion (O^{2-}) is unstable in the gas phase, because of electron–electron repulsion, spontaneously decomposing into O^- plus an electron. The oxide ion is stabilized in the bulk crystal by the electrostatic crystal field (the Madelung potential), but it is likely that, especially in small clusters, the oxygen species are not present as formal, rigid O^{2-} ions. Alternative approaches for circumventing this problem are—to introduce partial covalency (electron sharing) by reducing the formal ionic charges—or to include dipole (and possibly higher) polarization terms of the oxide ion, thereby acknowledging that the outer electron density of the O^{2-} ion is easily deformed in a way which depends on the number, and arrangement, of surrounding cations. The polarization of the oxide anion allows for partial screening of the repulsive interactions between cations, as well as lowering the energy of the oxide ions themselves. The polarizability of the much smaller Mg^{2+} cation is some three orders of magnitude smaller than that of O^{2-} and so can be ignored.

Ziemann and Castleman used a simple rigid ion model (a pair potential, as in Equation (7.1)) to predict the geometries of neutral, stoichiometric clusters $(MgO)_N$. They took two sets of formal ion charges: $q = \pm 1$ $(Mg^+O^-)_N$; and $q = \pm 2$ $(Mg^{2+}O^{2-})_N$ and found that the larger formal charges favoured open cage-like structures, with a high number of six-membered Mg_3O_3 and eight-membered Mg_4O_4 rings, in addition to Mg_2O_2 squares. These open structures result from the large repulsive interactions between ions with the same charge (cation-cation and anion-anion). The lower charges, which were introduced to model covalent effects (see above), lead to the formation of predominantly bulk-like cluster structures, composed of Mg_4O_4 cubes, where all rings are square. Polarizable ion model calculations (using formal charges of ± 2 on the ions but allowing for anion compressibility and polarization) by Wilson found that rock-salt fragments become favoured for $(MgO)_N$ clusters with $N \geq 30$. For the smaller clusters, however, Wilson's calculations indicated a preference for stacked hexagonal nanotube structures—consistent with the $(MgO)_{3n}$ magic numbers observed experimentally. Comparing these results with previous studies of alkali halide clusters (see Figure 7.4), the observation of this hexagonal structural motif means that neutral $(MgO)_N$ clusters are more closely related to LiF clusters than clusters of NaCl or KCl. Although the F^- anion is much less polarizable than Cl^-, the polarizing ability of the small Li^+ cation is significantly larger than for Na^+ or K^+.

Ab initio electronic structure calculations have been performed on neutral and charged MgO clusters, both with stoichiometric and non-stoichiometric compositions. Methods adopted include Hartree–Fock and post-Hartree–Fock MO calculations (where the latter include dynamic electron correlation) and DFT. The results for neutral $(MgO)_N$ clusters are in general agreement with those using polarisable ion models—with hexagonal structures giving way to bulk-like cubic structures with increasing cluster size. Some differences were observed, however, such as the prediction of the truncated octahedron as the most stable isomer of $(MgO)_{12}$—this structure is actually the lowest energy structure for the rigid ion model with charges $q = \pm 2$.

Ab initio calculations on non-stoichiometric clusters with excess Mg ions, $(MgO)_N Mg^{2+}$, have indicated that these clusters tend to adopt cubic structures, even at low nuclearities—as these clusters have an odd number of ions, complete cuboidal clusters (i,j,k) can be constructed when the product $i \times j \times k$ is odd, whereas hexagonal prismatic clusters must possess an even number of ions. It has also been shown that the excess charge in these non-stoichiometric (defect) clusters is screened more effectively in cubic clusters. An important prediction of these calculations is that the lowest energy structure for the 27-ion cluster $(MgO)_{13} Mg^{2+}$ is not the (3,3,3) cube (Figure 7.1), since this structure would possess a central (bulk-like) oxide anion which is octahedrally coordinated by Mg^{2+} ions. Although there is an electrostatic (Madelung) stabilization of oxide in this environment, such a configuration is higher in energy than structures in which the oxide ions are all on the surface of the cluster. Oxide ions in low-coordination (and low-symmetry) surface sites have greater polarisabilities, leading to a greater stabilizing polarization contribution to the cluster binding energy. Similar arguments explain the prediction that $(MgO)_{22}Mg^{2+}$ does not adopt the (3,3,5) cuboidal structure. As the clusters get larger, however, bulk-like structures including non-surface oxide ions do start to become more stable (from around $N = 24$ onwards).

7.5 METAL CHALCOGENIDE CLUSTERS

The solid compounds formed between the late transition metals (e.g. copper, zinc and cadmium) and the heavier chalcogenide (group 16) elements (sulfur, selenium and tellurium) exhibit bonding which is intermediate between covalent and ionic. The ionic character of the corresponding metal chalcogenide clusters will decrease as the difference in electronegativity between the metal and the non-metal atom increases. Considering the zinc and cadmium chalcogenides, since zinc is more electropositive than cadmium and the electronegativity of the chalcogenides varies as S > Se > Te, ZnS clusters have the most ionic character, while CdSe and CdTe clusters have the least ionic character. For convenience, in this section, I will adopt the ionic convention—thus in clusters $(MS)_N$ ($M = $ Zn, Cd) I will talk about formal ionic charges M^{2+} and S^{2-}, while recognizing that there may be a significant amount of covalency in the bonding of these clusters.

7.5.1 Zinc Sulfide Clusters

Heating solid ZnS in a vacuum produces a vapour which consists primarily of Zn atoms and S_2 molecules. Mass spectroscopic studies by Martin and co-workers have detected an extensive series of clusters, generated by quenching the vapour above ZnS in helium. As shown in Figure 7.5, the detected clusters, which are ionized so that they may be detected in the mass spectrometer, are either stoichiometric $(ZnS)_N^+$, or else have an excess S atom, $(ZnS)_N S^+$. In the non-stoichiometric clusters, one of the sulfide (S^{2-}) anions of the stoichiometric cluster is replaced by a disulfide (S_2^{2-}) anion, so these clusters may alternatively be written as: $[Zn_N S_{N-1}(S_2)]^+$. Particularly intense peaks are observed for $(ZnS)_{13}$, $(ZnS)_{33}$, and $(ZnS)_{34}$.

Martin has proposed plausible structures for $(ZnS)_N$ clusters, as fragments of the cubic zinc blende structure of bulk ZnS, which is based on the diamond structure. The zinc blende structure can be decomposed into two interpenetrating fcc lattices, one of Zn^{2+} cations and the other of S^{2-} anions. Removing a face from an octahedron cut out of the ZnS lattice (so as to maintain the 1:1 stoichiometry) generates a series of $(ZnS)_N$ clusters, with a magic number (N^*) of ZnS units given by:

$$N^* = \frac{1}{6} K \left(4K^2 - 3K - 1 \right) \tag{7.5}$$

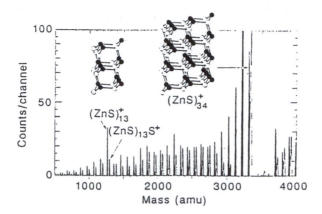

Figure 7.5 Mass spectrum of ZnS clusters, showing structures proposed to explain large peaks in the spectrum.

where K is the number of Zn^{2+} ions along each edge of the octahedron. The number of ZnS units corresponding to $K = 3, 4, 5$ and 6 are $N^* = 13, 34, 70, 125$. The octahedral zinc blende type structures proposed for $(ZnS)_{13}$ and $(ZnS)_{34}$ are shown in Figure 7.5. As mentioned above, the magic numbers $N^* = 13$ and 34 correspond to intense peaks in the $(ZnS)_N$ mass spectrum. Although there is no enhanced peak intensity for $N = 70$, there is a strong peak corresponding to $(ZnS)_{124}$ and a marked drop in intensity at $(ZnS)_{126}$.

7.5.2 Cadmium Sulfide and Selenide Clusters

Single crystal X-ray diffraction measurements on colloidal CdS particles (see Chapter 8) have shown that the cluster $Cd_{32}S_{55}$ has a structure which is a small fragment of the cubic bulk CdS (zinc blende type) lattice. TEM studies of surface-deposited colloidal CdSe particles, however, have shown that CdSe clusters grow in a hexagonal fashion, consistent with the hexagonal wurtzite structure of bulk CdSe. (Wurtzite is another allotrope of ZnS, which is based on the hexagonal diamond structure.) Bulk CdSe is known to undergo a phase transition from the wurtzite phase to the denser rock-salt (NaCl-type) structure at a pressure of around 2 GPa. Experimental studies on CdSe particles have shown that the wurtzite $\rightarrow$ rock-salt transition pressure increases with decreasing cluster size, showing an approximately linear dependence on $1/R$, as can be seen from Figure 7.6.

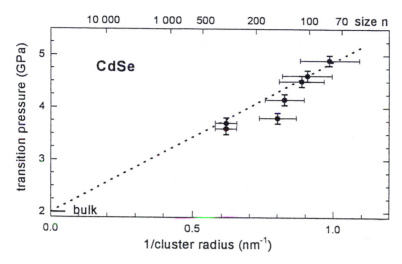

Figure 7.6 Size dependence of the transition pressure from the wurtzite to the rock-salt structure.

Bulk CdS and CdSe have fairly narrow band gaps and, as such, may alternatively be regarded as polar compound semiconductors (see Section 6.5). Spectroscopic measurements have shown that the HOMO–LUMO gap in CdSe clusters decreases with increasing size, tending towards the bulk band gap for large clusters. CdS and CdSe clusters are discussed further in Chapter 8, in the context of colloidal semiconductor particles and quantum dots.

7.6 METAL–CARBON CLUSTERS

In the final section of this chapter, I will discuss a recently initiated area of cluster chemistry—that of metal–carbon clusters. As will become apparent, some of these clusters are best regarded as covalently bonded semiconductor-like clusters (similar to those discussed in Chapter 6) while others have identical structures to the ionic alkali metal halide clusters.

7.6.1 Metallocarbohedrene Clusters: Met-cars

In 1992, Castleman and co-workers reported a very intense feature in the mass spectrum recorded after the laser vaporization of titanium metal in the presence of methane gas. The magic number corresponded to a cluster with the formula Ti_8C_{12} (though the cation was detected in the mass spectrometer). The occurrence of one dominant peak, reminiscent of the case of C_{60} discussed above, has a led a number of researchers to investigate this molecule and analogous clusters (which are formed by a large number of early transition metals), which were named metallocarbohedrenes, or 'met-cars' for short and which are regarded as organometallic equivalents of the fullerenes.

7.6.1.1 Generation of Met-cars

Castleman used a laser-induced plasma reactor, based on the methods of Smalley and Bonybey for producing clusters of refractory metals and non-metals (e.g. Si), wherein the metal is vaporized by a high power pulsed laser and the metallic vapour is subsequently cooled in a high pressure carrier gas stream. For generating met-cars, gaseous hydrocarbons (especially methane) are used as the carrier gas, either alone or in combination with helium. Using this apparatus, the Castleman group synthesized met-cars, M_8C_{12}, of the group 4 and 5 transition metals Ti, Zr, Hf, V and Nb. Macroscopic quantities of the Ti and V met-cars have been produced. Pilgrim and Duncan used the same technique to generate met-cars of Cr, Mo and Fe. Met-cars have also been generated by laser co-vaporization of metal and graphite and a series of mixed-metal met-cars with compositions $Ti_{8-x}Zr_xC_{12}$ have been produced by laser vaporization of a mixture of TiC and Zr powders. Other mixed-metcars $Ti_{8-x}M_xC_{12}$ (M = Y, Nb, Ta, Mo, W, Si) have been generated using similar methods. Pulsed arc cluster ion sources, similar to those employed to generate fullerenes, have been used to synthesize met-cars. Finally, met-cars are often generated as products of the photo-fragmentation of larger species, such as the transition metal-coated fullerenes $C_{60}M_x$ and $C_{70}M_x$ (M= Ti, V)—see Section 6.2.2.9.

In addition to the very intense peaks corresponding to met-cars, mass spectroscopy also reveals the presence of smaller fragments, such as $[M_4C_8]^+$, $[M_5C_{10}]^+$, $[M_6C_{12}]^+$ and $[M_7C_{13}]^+$ (M = Ti, V). For smaller sizes, the most abundant species are $[MC_2]^+$, $[M_2C_4]^+$ and $[M_3C_6]^+$. These results are consistent with met-car formation via the initial condensation of MC_2 groups, followed by metal addition.

7.6.1.2 Structures of Met-cars

The widespread occurrence of met-cars and their high relative abundance implies that the M_8C_{12} cluster is particularly stable. A number of structures, generally invoking high symmetry, have been proposed to account for this stability.

Castleman initially proposed that met-cars have dodecahedral structures, comprising pentagonal M–C–C–M–C rings. This structure, which is shown in Figure 7.7a, is a heteronuclear equivalent of the smallest fullerene, C_{20} and has overall T_h symmetry. The dodecahedral structure can be constructed as a (non-bonded) cube of metal atoms, with C_2 units capping each face of the cube. In this model, all of the metal atoms are equivalent by symmetry. Pauling suggested a similar structure, in which the C_2 units lie within the faces of the cube, rather than above them, but this was subsequently calculated to be significantly less stable than the dodecahedral structure.

On the basis of DFT calculations, Dance proposed an alternative structure, which has been gaining support, from theoretical and experimental studies and which is know generally held to be the true structure of the met-car cluster. In this model, shown in Figure 7.7b, the met-car structure consists of a tetracapped tetrahedron of metal atoms, with C_2 units bridging the edges of the tetrahedron. A variety of *ab initio* (Hartree–Fock and correlated) and DFT calculations indicate that the tetracapped tetrahedral structure is more stable than the dodecahedral structure by 800–1,470 kJ mol^{-1}. A detailed DFT study by Bénard and colleagues has shown that the tetracapped tetrahedral structure, which has overall T_d symmetry and two sets of four symmetry-equivalent metal atoms, is stable with respect to symmetry-lowering distortions.

An important stabilizing contribution to the bonding in met-cars is the back-donation of electronic charge from filled metal d orbitals into empty C–C antibonding π^* orbitals on the bridging C_2 units. (This interaction is analogous to the back-bonding interaction responsible for stabilizing metal complexes of unsaturated organic ligands, such as ethene (ethylene) and ethyne (acetylene).) The back-bonding also favours the tetracapped tetrahedral structure over Castleman's dodecahedral structure. Interestingly, a number of crystalline chalcogenide compounds of copper have been found to contain dodecahedral are Cu_8E_{12} clusters (E = S, Te). Back-bonding is unimportant in these compounds, since the S_2 or Te_2 units do not have any empty π^* orbitals.

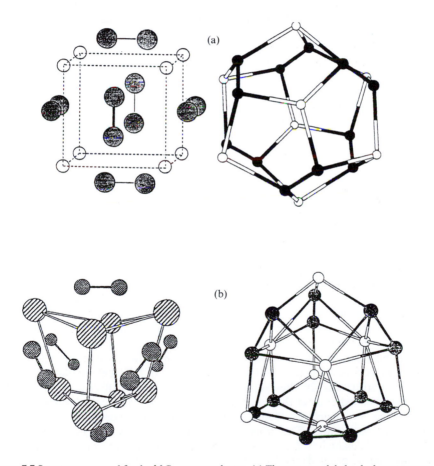

Figure 7.7 Structures proposed for the M_8C_{12} met-car cluster. (a) The pentagonal dodecahedron structure, with eight equivalent metal atoms (T_h symmetry). (b) The tetracapped tetrahedron structure, with two sets of four equivalent metal atoms (T_d symmetry).

7.6.1.3 Experimental Studies of Met-cars

Ion mobilities measured for $[Ti_8C_{12}]^+$ by Bowers and co-workers, in drift tube experiments, have been compared with calculated mobilities for a variety of met-car models. Although the best agreement was found for the dodecahedral structure, the tetracapped tetrahedral structure also showed a reasonably good fit to the experimental data. The ionization energy (4.40 ± 0.02 eV) measured for Ti_8C_{12} is in excellent agreement with that calculated by Dance (4.37 eV) for the tetracapped tetrahedral structure. The low ionization energy, coupled with the low electron affinity (the adiabatic electron affinity is 1.05 ± 0.05 eV) is consistent with calculations which indicate the highest occupied orbitals of Ti_8C_{12} to be non-bonding in nature.

Photodissociation mass spectra of $[M_8C_{12}]^+$ ions (M = Ti, V) show fragment ions $[M_xC_{12}]^+$ (x = 5–7) which indicate loss of metal atoms only. In the case of $[Zr_8C_{12}]^+$, however, initial loss of a Zr atom appears to be followed by loss of ZrC_2 units, resulting in the fragmentation products $[Zr_7C_{12}]^+$, $[Zr_6C_{10}]^+$, $[Zr_5C_8]^+$, and $[Zr_4C_6]^+$.

The chemical reactivity of met-car clusters has been investigated by Castleman and by Byun and Freiser, by passing a mass-selected beam of clusters through a drift tube containing reactant molecules in a carrier gas. $[Ti_8C_{12}]^+$ was found to be very reactive towards polar molecules (Mol = H_2O, NH_3, CH_3OH), giving rise to association products $[Ti_8C_{12}]^+(Mol)_n$ (with $1 \leq n \leq 8$). Although the uptake of up to eight molecules has been said to confirm the dodecahedral met-car structure (with eight equivalent metal sites), Bénard and co-workers have calculated that the same pattern should be observed for the tetracapped tetrahedral structure. The T_d structure also explains the large MS peak observed corresponding to four absorbed methanol molecules and the fact that in the case of π-bonding ligands (such as benzene and ethylene), which bind more weakly to the cluster, a maximum of four ligands can be bound to the cluster. Finally, complexes have been detected between methane and Ti_8C_{12}, with the dominant peak corresponding to $Ti_8C_{12}(CH_4)_4$.

7.6.2 Copper Carbohedrene Clusters

In 1993, Yamada and Castleman performed experiments in which pre-formed copper clusters were allowed to react with acetylene in the gas phases, before cooling, laser ionization and subsequent mass spectral analysis. They detected a series of copper carbohedrene ions $[Cu_xC_y]^+$, with particularly prominent peaks for the series $[Cu_{2N+1}C_{2N}]^+$ (N = 1–10). On the basis of DFT calculations, Dance has proposed structures for these clusters which are quite different for those of the metallocarbohedrenes of the early transition metals. The calculations indicate a preference for structures involving a close-packed copper core with C_2 units bonded to its surface. The proposed structure for $[Cu_{13}C_{12}]^+$, for example, has a centred cuboctahedral Cu_{13} core, with each of the six square faces of the cuboctahedron bridged by a C_2 group.

7.6.3 Nanocrystalline Metal Carbide Clusters

In the course of repeating the experiments of Castleman and co-workers on Ti met-cars, Pilgrim and Duncan observed a series of intense peaks in their mass spectra at masses corresponding to larger clusters than $[Ti_8C_{12}]^+$. Similar series have been found for other metals, such as V and Zr. These larger magic number clusters, which have M:C ratios close to 1:1, have been attributed to nanocrystalline cuboidal clusters, which are fragments of the bulk MC lattice—which has the fcc rock-salt (NaCl) structure. This description of the clusters as carbide-like is consistent with the fact that their IR spectra show no C–C stretching vibrations.

The intense peak corresponding to $[M_{14}C_{13}]^+$ (M = Ti, Zr, V) has been assumed to have the same cubic (3,3,3) structure previously encountered for the ionic $M_{14}O_{13}$ and $M_{14}X_{13}$ (X = halide) clusters. Similarly, the peak at $[M_{18}C_{18}]^+$ has been attributed to a (3,3,4) cuboidal geometry. Magic number cuboidal clusters have been observed up as far as the (5,5,5) cube. For the cubic clusters $[M_{14}C_{13}]^+$ (3,3,3) and $[M_{63}C_{62}]^+$ (5,5,5), the mass spectral maxima correspond to the perfect cubes, as these structures have metal atoms at the eight corners of the cube. By contrast, for cuboidal clusters (i,j,k), with the product $i{\times}j{\times}k$ = even, loss of carbon atoms is observed from the corner sites. This is opposite to the behaviour observed for MgO clusters, where there is a preference for the oxide anions to occupy the low-coordinate surface, edge and corner sites in cuboidal clusters.

Pilgrim and Duncan have measured the photodissociation mass spectra of the nanocrystalline metal carbide clusters. They found that photodissociation of the cubic $[Ti_{14}C_{13}]^+$ cluster leads to loss of exclusively metal atoms, with the dominant dissociation channel being:

$$[Ti_{14}C_{13}]^+ + h\nu \rightarrow [Ti_8C_{13}]^+ + 6Ti$$

The major photofragment is thus $[Ti_8C_{13}]^+$, rather than the met-car ion $[Ti_8C_{12}]^+$. It has been postulated that the extra carbon atom resides in the centre of a met-car cage, so that $[Ti_8C_{13}]^+$ is actually an endohedral met-car, which by analogy with endohedral fullerene complexes may be written as $[C@Ti_8C_{12}]^+$. Similar results are obtained for V, though for Zr the main photofragment is the met-car $[Zr_8C_{12}]^+$.

For the larger (3,3,4) clusters $[M_{18}C_{18}]^+$, photodissociation appears to involve the cleaving off of one face of the nanocrystal, leading to the (3,3,3) cube and a nine-atom cluster, either of which can carry the positive charge:

$$[M_{18}C_{18}]^+ + h\nu \rightarrow [M_{14}C_{13}]^{(+)} + [M_4C_5]^{(+)}$$

7.7 EXERCISES

7.1   Why does the '6×4×3' cuboidal isomer of $(NaCl)_{35}Cl^-$ have a shorter
        drift time in ion mobility studies than the '8×3×3' isomer? What would
        be the relative mobility of a '6×6×2' isomer in comparison?

7.2   What experiments can be used to obtain structural information on ionic
        clusters?

7.8 FURTHER READING

Beck, R.D., St. John, P.M., Homer, M.L. and Whetten, R.L., 1991, Impact-
    induced cleaving and melting of alkali-halide nanocrystals. *Science*, **253**, pp.
    879–883.
Martin, T.P., 1983, Alkali-halide clusters and microcrystals. *Physics Reports*, **95**,
    pp. 167–199.
Martin, T.P., 1994, Oxides and halides of alkali metals. In *Clusters of Atoms and
    Molecules*, Vol.I, edited by Haberland, H., (Berlin, Springer-Verlag), pp. 357–
    373.
Martin, T.P., 1996, Shells of atoms. *Physics Reports*, **273**, pp. 199–241.
Rohmer, M.-M., Bénard, M. and Poblet, J.-M., 1999, Metallocarbohedrenes
    M_8C_{12} and $Ti_xM'_yC_{12}$ – from mass spectrometry to computational chemistry. In
    Metal Clusters in Chemistry, Vol.3, edited by Braunstein, P., Oro, L.A. and
    Raithby, P.R., (Weinheim, Wiley-VCH), pp. 1664–1710.
Rohmer, M.-M., Bénard, M. and Poblet, J.-M., 2000, Structure, reactivity and
    growth pathways of metallocarbohedrenes M_8C_{12} and transition metal/carbon
    clusters and nanocrystals: a challenge to computational chemistry. *Chemical
    Reviews*, **100**, pp. 495–542.
Whetten, R.L., 1993, Alkali halide nanocrystals. *Accounts of Chemical Research*,
    26, pp. 49–56.

CHAPTER EIGHT

Clusters in Action: Past, Present and Future

8.1 INTRODUCTION

In this chapter, I will introduce a number of topics, such as colloidal metal particles, semi-conductor quantum dots and cluster deposition, which have already led to technological applications of clusters and which promise to open up whole new avenues of nanotechnology in the future. Many more actual and possible applications have also been discussed in previous chapters, when talking about specific types of clusters.

8.2 COLLOIDAL METAL PARTTICLES

8.2.1 Background

One problem with studying naked metal clusters (such as those created in cluster molecular beams) is that they cannot be isolated and handled on a preparative scale like conventional molecules. To enable the investigation of approximately uniformly sized clusters, and to exploit cluster properties in device applications, it is necessary to protect ('passivate') them with a ligand shell, as this avoids coalescence at high cluster densities. For clusters of a few atoms or tens of atoms this leads to a wealth of inorganic metal cluster compounds, which have been studied extensively over the past 30–40 years.

In the larger size regime of clusters ranging from nanometre to micron dimensions, lie the old and venerable class of metal 'colloids', some of which have been known for many centuries. For example stained glass windows, dating back as far as medieval times, provide beautiful examples of colours which are caused by the suspension of small colloidal particles of copper, silver and gold in the glass. In fact, the use of colloidal metals in colouring glass probably dates back as far as the ancient Egyptians—Cleopatra is reported to have used cosmetics prepared from (presumably colloidal) gold.

A colloid is a dispersion of particles (generally sub-micron in size) of one material in another. A dispersion of a solid in a liquid (e.g. a passivated metal cluster in water or an organic solvent) or a solid dispersed in another solid (such as in the stained glass mentioned above) is known as a 'sol'. For a discussion of the structures of large colloidal particles, see Section 4.6. Metal–insulator transitions in single isolated and arrays of colloidal metallic particles have been discussed in Section 5.3.

8.2.2 Preparation of Metallic Colloids

Colloidal metallic particles are generally produced by chemical reduction of metal salts dissolved in an appropriate solvent, in the presence of a surfactant (e.g. citrate, alklythiols or thioethers) or polymeric ligands (e.g. polyvinylpyrrolidone), which passivates the cluster surface. A colloidal metal particle can be described in terms of a metallic core surrounded by a ligand shell, as shown in Figure 8.1. Alkylthiols $(CH_3(CH_2)_nSH)$ and thioethers $(CH_3(CH_2)_nS(CH_2)_mCH_3)$ form particularly stable colloidal particles with metals (M = Ag, Au, Pd, Pt) due to the strength of the M–S bond (especially for gold).

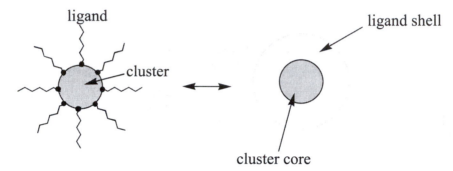

Figure 8.1 Schematic representation of a colloidal metal particle, consisting of a metallic core surrounded by a ligand shell.

Colloid preparation can be single-phase or two-phase, where colloids form at the interface between an aqueous layer (in which the metal salt is dissolved) and an organic layer (containing the surfactant and reducing agent) and they can also be generated by using inverse micelles. (An inverse micelle is a hollow, spherical species, the inner surface of which is made chemically attractive to metal ions.) Control of average particle size and size-distribution can be achieved by tuning the preparation conditions (solvent, surfactant, reducing agent, etc.), or varying the size of the inverse micelles.

Bimetallic ('nanoalloy') colloids (e.g. Ag–Pd and Cu–Pd) can be prepared by chemical reduction of the appropriate mixture of salts in the solution phase. As mentioned in Section 5.2, Schmid and co-workers have prepared layered (shell-like) colloidal Pd-covered Au clusters (Au/Pd) and Au-covered Pd clusters (Pd/Au) by the 'seed-germ' process, wherein a gold (or palladium) cluster without passivating ligands undergoes a second reduction step in the presence of salts of the other metal. The outer metal shell is then passivated by coordination of an appropriate sheath of ligands.

Figure 8.2 HRTEM image of shell structured Pd-coated Au clusters. The 18 nm Au cores (dark areas) are covered by a Pd shell of thickness 4–5 nm.

Figure 8.2 shows a high resolution transmission electron micrograph (HRTEM) of shell structured Pd-coated Au clusters, where the heavier Au core (of diameter 18 nm) shows up darker than the outer shell (of thickness 4–5 nm) of Pd.

8.2.3 Colours of Metallic Coloids

The scientific study of metallic colloids dates back to 1857, when the polymath Michael Faraday made the remarkable observation that in gold colloids: '... the gold is reduced in exceedingly fine particles which becoming diffused, produce a beautiful fluid ... the various preparations of gold whether ruby, green, violet or blue ... consist of that substance in a metallic divided state'. We now understand the colours of colloidal suspensions or 'sols' in terms of strong absorption bands in the visible region of the spectrum caused by plasmons—i.e. collective oscillations of the cluster electrons (see Section 4.2.3). The colours of colloidal metals have been exploited for many centuries, with stained glass windows (consisting of colloidal copper, silver and gold clusters in a glass matrix) being particularly beautiful examples.

According to Mie theory (Equation (4.5a)), the plasmon frequency should decrease as the cluster radius increases. The absorption characteristics of colloidal metal particles are, however, very complex—depending on the electronic structure of the metal, the size of the cluster, the type of plasmon (conduction vs. valence electrons, dipole vs. quadrupolar, etc.), the dielectric constant of the medium (matrix) in which the metal particles are suspended/embedded and the nature of the cluster surface–matrix interaction. At higher concentrations of colloidal particles (e.g. in particle arrays), there may also be interference effects—

i.e. coupling between the individual particles (see Section 5.3.3). Other light scattering effects, such as opalescence, can also be observed.

8.2.4 Structures of Colloidal Metal Particles

The study of large colloidal particles using electron microscopy has already been discussed (see Section 4.6 and the micrograph shown in Figure 4.15). Schmid and co-workers have reported the synthesis and characterization of a number of smaller passivated clusters with approximately 10–1,000 atoms. Many of these clusters, such as Au_{55}, Pt_{309}, Pd_{561} and Pd_{1415} have fcc-like cuboctahedral structures. In terms of the geometric shell model (see Chapter 4), these correspond to two-, three- five- and six-shell clusters, respectively. The one-shell Au_{13} cluster, however, has an icosahedral structure. The non-icosahedral nature of the larger clusters (by contrast to bare metal clusters) is probably due to structural changes induced by ligand coordination.

In fact, only for the smallest of these clusters (Au_{13}) has the structure been determined by single crystal X-ray diffraction. The structures of the larger clusters have been determined (or inferred) using a variety of techniques, including electron microscopy, electron diffraction, small-angle X-ray scattering, Mössbauer spectroscopy and extended X-ray absorption fine structure spectroscopy (EXAFS). For example, EXAFS measurements of nearest-neighbour coordination numbers have been compared with analytically derived mean coordination numbers to estimate the size of certain ligand-stabilized metal particles. ^{197}Au Mössbauer spectroscopy reveals that both the surface atoms and the inner atoms of Au_{55} clusters are influenced by the ligands and that the central atom charge densities are not the same as those in the bulk metal. By contrast, in Pt_{309} clusters (in which ^{197}Au nuclei are produced by neutron activation) the inner shell atoms have the same charge density as in the bulk metal.

8.2.5 Colloidal Crystals

Exciting advances have been made (by a number of research groups around the world) in producing colloidal suspensions (sols) with narrrow size distributions and in the fractional precipitation of passivated metal clusters from a sol to generate colloidal crystals or 'opals'. These solids may be considered as superlattices, which are fascinating examples of materials that are ordered on two different length scales, since there is a regular arrangement of atoms within the clusters (with spacings of the order of 0.3 nm), while the clusters themselves form a regular 3-D arrangement, with spacings of a few nanometres (depending on the size of the passivating ligands). Whetten and co-workers have used X-ray diffraction to solve the structure of a crystal of size-selected truncated-octahedral Au_{38} clusters, coated by alkylthiol ligands. A representation of a superlattice of passivated gold clusters is shown in Figure 8.3.

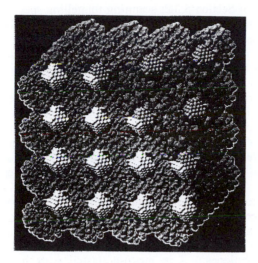

Figure 8.3 Schematic representation of a colloidal crystal—a superlattice of passivated gold clusters.

Schiffrin and colleagues found that when they added decanethiol ($C_{10}H_{21}SH$) to a sol consisting of gold particles (with diameters $D \approx 7$ nm) stabilized by $R_4N^+Br^-$ salts, in toluene, some of the clusters were etched, resulting in a bimodal cluster size distribution. Denoting the two differently sized clusters as A and B (with $D_A = 7.8 \pm 0.9$ nm and $D_B = 4.5 \pm 0.8$ nm), the ratio of the diameters (and hence the radii) of these particles $D_B/D_A \approx 0.58$ is ideal for forming an ordered hexagonal AB_2 array—which is indeed what is observed! TEM studies of these phases show a number of hexagonal AB_2 rafts, with areas of around 1 μm^2. Interestingly, natural opals are sols which crystallize in such an AB_2 arrangement.

8.2.6 Applications of Metal Colloids

The ability to generate metal colloids with narrow size distributions (using chromatography and centrifuge techniques, as developed by Whetten and colleagues) means that it is now possible to design arrays of nanoparticles, with dimensionality ranging from zero (isolated clusters) to three (as in colloidal crystals).

8.2.6.1 Isolated Particles

The passage of electric current through bulk metal occurs due to the motion of a large number of free electrons, moving under the influence of the applied potential. The current flow is averaged over all the charge carriers and is continuous, so that the resistance obeys Ohm's law. If, however, a single metallic cluster (nanoparticle) is placed between two electrodes, the number of carrier electrons is much smaller and can be controlled by adjusting the voltage. One can, therefore, consider the motion of single electrons via metal nanoparticles.

As mentioned in Section 5.3.2, scanning tunnelling spectroscopy (STS) has been used to study the *I–V* characteristics of individual colloidal metal particles (e.g. passivated Pt_{309} and Au_{55} clusters on gold substrates). The passivating organic ligands are insulating and act as tunnel barriers between the cluster and the substrate. This results in a 'Coulomb blockade', so that a critical charging voltage (V_c), corresponding to a charging energy ($E_c = eV_c$), is required for electrons to tunnel between the STS tip and the cluster and between the cluster and the substrate. For passivated Pt_{309} clusters on Au, Schmid and coworkers have measured charging energies ranging from 50–500 meV. Andres and colleagues used STS to measure the cluster-substrate capacitance (1.7×10^{-17} F) of 1.8 nm Au particles grown on a dithiol modified Au surface. This small capacitance enables the observation of Coulomb blockade phenomena at room temperature.

The tuneable capacitance and resistance of metal cluster particles allows the manipulation of electrons at the single-electron level—thereby opening up the possibility of fabricating nanoscale circuits and devices such as single electron transistors. Such circuits could be constructed by deposition and/or nanolithographic techniques (e.g. by moving clusters with the tip of an STM or AFM) of arrays of nanoparticles.

A *single electron tunnelling* (SET)-based transistor has been reported by Sato *et al.* Metal electrodes (source, drain and gate) were fabricated on an SiO_2 support, using electron beam lithography. After deposition of citrate-stabilized colloidal Au particles and subsequent replacement of the citrate molecules by hexanedithiol spacer ligands, it was found—using scanning electron microscopy (SEM)—that a chain of three linked Au clusters bridged the source and drain electrodes, as shown in Figure 8.4. Single electron charging, as evidenced by a Coulomb gap, dominates the electronic conduction for temperatures up to 77 K.

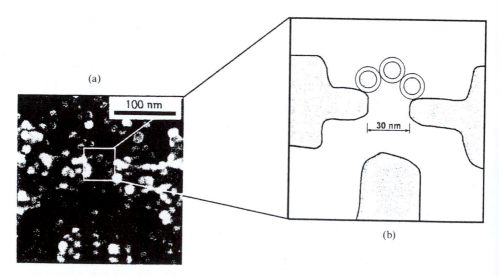

Figure 8.4 (a) SEM image of a chain of three gold nanoparticles incorporated into a system of source, drain and gate metal electrodes. (b) Schematic diagram of the nanoparticles and electrodes.

Turning to non-metallic particles, Alivisatos and colleagues have fabricated a single electron transistor from a 5.5 nm CdSe nanoparticle bound to two closely spaced ($\approx$ 5 nm) gold electrodes using bifunctional ligands. Recent work has also been reported on the use of single C_{60} clusters as transistors in nanoelectronic devices.

8.2.6.2 One-Dimensional Arrays

Schmid and co-workers have used the ordered channels of porous alumina (with a diameter of 7 nm) as templates to obtain linear chains of ligand-stabilized Au_{55} clusters. In principle, the diameter of the resulting 'nanowire' can be controlled by varying the pore width, although defects and disorder in the chains is a problem which is still to be overcome. In particular, the particle diameter distribution should be narrow as the presence of small clusters may lead to electron localization (trapping)—i.e. a weak link in the chain. In an interesting recent development, Alivisatos and co-workers have generated linear chains by coordinating Au clusters ($D \approx 1.4$ nm) to a single strand DNA oligonucleotide of specific length and sequence.

8.2.6.3 Two-Dimensional Arrays

In the early 1990s, Schiffrin and co-workers produced gold organosols in organic solvents, using alkane thiols as surfactants. These nanoparticles readily self-assemble into 2-D arrays on solid supports, such as highly oriented pyrolized graphite (HOPG), yielding close-packed (hexagonal) structures extending over tens or hundreds of nm, with separations of approximately 1 nm between particles (see also Section 8.2.5). X-ray diffraction yields a low-angle peak corresponding to a lattice plane spacing of 5 nm, indicating that there is some overlap between the ligand spheres of the clusters, presumably due to interdigitation of the ligands.

The TEM image of a thiol-derivatized gold nanoparticle array deposited on graphite (produced by Rao and co-workers) is reproduced in Figure 8.5a. The particles are nearly spherical with a mean diameter of ca. 4 nm, as shown in Figure 8.5b. The gold nanoparticles assemble into a hexagonal superstructure, (as shown schematically, earlier in Figure 5.9) and the distance between the particles is nearly constant, at approximately 0.55 nm, throughout the superstructure domain. In a study of arrays of colloidal Pd particles, passivated by alkane thiols of varying lengths, Rao and co-workers found that well ordered 2-D arrays were observed when the ratio of particle diameter (D) to thiol chain length (ℓ) lies in the range 1–4, with the most highly ordered films resulting for D/ℓ ratios of 2–3.

Thin films of passivated colloidal metals show great promise for making new types of electronic and optical devices. For example (as discussed in Section 5.4), Heath and co-workers have demonstrated reversible pressure-induced metallization in a Langmuir film of thiol-coated silver clusters, which may lead to a nanoscale electrical switch. Andres and colleagues have reported charge transfer through a hexagonal 2-D network of 4 nm Au particles connected by rigid conjugated organic molecules (spacer ligands), resulting in a cluster spacing of 2.2 nm.

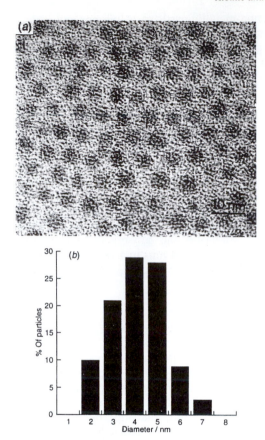

Figure 8.5 (a) TEM image of a thin film of thiol-coated gold particles. (b) Histogram showing the size distribution of the gold particles.

8.2.6.4 Three-Dimensional Arrays

Depositing multiple layers of thiol-coated clusters leads to multilayers which are fragile, as the inter-layer bonding is due to relatively weak van der Waals interactions between the hydrophobic alkyl chains. More rigid multilayer superstructures have been constructed by alternately dipping a substrate (e.g. a solid gold surface) in a solution of a dithiol (an alkyl chain with a thiol group at either end: $HS(CH_2)_nSH$, which can thus bond to two metal particles—acting as a bridging ligand) and a sol of clusters passivated by loosely bound ligands, which are readily displaced by thiols. In this way, 3-D structures can be built up layer by layer.

The Schmid group have studied 3-D arrays of passivated Pd_{561} clusters, with bifunctional amine molecules ($H_2N–\square–NH_2$) as spacers (where $\square$ symbolizes a generic alkyl or aryl chain). They found that variation of the length and nature of the spacer ligands presents a new method for fine tuning the properties of a nanoparticle array—thus increasing the length of the spacer ligands leads to an increase in the charging energy, while conjugation reduces it. Schiffrin and co-workers have obtained similar results for 2.2 nm and 8.8 nm colloidal Au nanoparticles interconnected by alkyldithiols of varying lengths.

8.2.6.5 Further Considerations

The layer-by-layer cluster deposition method, utilizing bifunctional ligands, also enables the construction of hetero-arrays, with alternating layers of two different metal clusters (e.g. Au_N and Pt_M), or even alternating layers of metal and semiconductor particles (e.g. Au_N and $(CdS)_M$), as shown schematically in Figure 8.6a.

As well as building up layer structures, dithiols can also be used to generate cross-linked 2- and 3-D arrays of metal particles. Additionally, bifunctional ligands (X–☐–Y) with different functional groups on each end (X, Y = SH, OH, NH_2 etc.) can also be employed to form ordered 1-, 2-and 3-D hetero-arrays of clusters of different metals (A_a and B_b), provided that (for example) functional group X binds more strongly to metal A than to B, the reverse being true for group Y. Such an arrangement is shown schematically in Figure 8.6b.

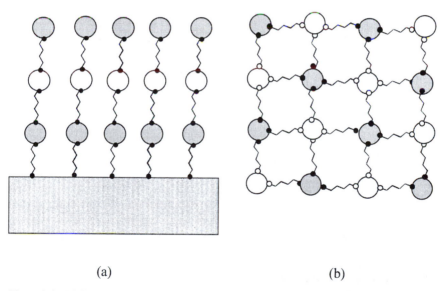

(a)   (b)

Figure 8.6 (a) Schematic representation of a hetero-layer arrangement of colloidal particles. (b) Schematic representation of an ordered hetero-array of colloidal particles linked by asymmetric bifunctional ligands.

Recent developments in colloidal metal clusters have included: the coordination of redox-active groups (e.g. ferrocene) to the ligand shell of a colloidal particle, in order to study the effect of the metal cluster core on the redox properties of the the tethered molecule. A number of electrochemical experiments have been performed using metal electrodes whose surfaces have been modified by tethering metal particles to them. As mentioned above, electron transfer between ligands can be studied by using electroactive spacers (bridging ligands), containing aromatic groups and/or conjugated unsaturated ($-CH=CH-$)$_n$ chains.

There are many possible applications of single electron tunnelling (SET) devices based on colloidal metallic nanoparticles—either isolated (0-D) or in 1-, 2- or 3-D arrays, as shown schematically in Figure 8.7. The 0-D SET transistor has been discussed above, other possibilities include the 1-D SET turnstile and 2-D SET quantum current transformer. In order to fabricate these (and other, as yet unimagined) devices, it will be necessary to develop better techniques for cluster size control and deposition/lithography, so that defects are not introduced during fabrication.

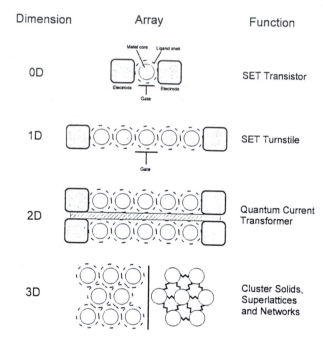

Figure 8.7 Possible SET arrangements and devices based on ligand-stabilized metal clusters.

8.3 SEMICONDUCTOR QUANTUM DOTS

8.3.1 Introduction

The development of the semiconductor industry was one of the defining characteristics of the latter half of the twentieth century. At the beginning of the twenty-first century, computers are developing in a number of directions— miniaturization, increased memory capacity and increased processor and communication speed. Nanocluster technology has the potential to impact substantially on all of these areas.

We are all familiar with bulk semiconductors, which have filled valence bands and empty conduction bands and a characteristic band gap which may lie in the visible or ultra violet regions of the spectrum. Doping with acceptor or donor atoms introduces impurity states in the band gap and the occurrence of thermally activated hole or electron conduction. As discussed in Chapter 6, there has been considerable interest in generating homonuclear clusters of semiconducting elements (such as silicon or germanium) and heteronuclear clusters based on II–VI and III–V compounds (such as CdS and GaAs) in order to see how their properties compare with those of the bulk and how they vary with size.

8.3.2 Quantum Dots

Studies of semiconductor nanoclusters have shown that the electron energy levels of the cluster, which can be described by a simple particle in a box model, are more widely spaced in small clusters, so that the band gap (the energy separation between the highest filled electron energy levels and the lowest empty levels) increases as the clusters get smaller. For cadmium sulfide, for example, the band gap increases from 2.5 eV in the bulk to 4.5 eV for small $(CdS)_N$ clusters. The confinement of the electrons within a 'zero-dimensional' particle, and the discrete nature of the electron energy levels of these particles, has led to the description of semiconductor nanoclusters as 'quantum dots'.

Quantum dots, which promise great sensitivity for measuring currents and local fields, may eventually be used as single-electron switching devices or transistors. Their small size (nanometre dimensions) means that a large number of quantum dots can be packed onto a computer chip. One problem which must be addressed is to obtain reproducible, narrow size-range clusters. Chemists are currently making significant progress along these lines, using chemical techniques and fractional crystallization methods.

Semiconductor nanoclusters may also be of use in optical computers, where digital information is transmitted as light pulses travelling along optical fibres, as they open up the possibility of manufacturing very small, highly efficient, quantum dot lasers which operate at low power, but which can produce laser light of high frequencies. Also, such lasers should be tuneable, since the light emitted will depend on the band gap, and hence the size of the component nanoclusters (see Section 8.4).

The development of devices based on quantum dots depends on being able to fabricate 1-, 2- and 3-D arrays of semiconductor nanoclusters which are connected to each other (so that information can pass between them in the form of electrons) but at the same time they are protected from coalescence. One way of achieving this has been pioneered by Heath and co-workers, who have fabricated a 'nanoscale egg carton' by using electron-beam lithography to carve parallel rows of 100 nm wide wells in a silicon wafer. Ge clusters are formed in these wells by condensation from the vapour phase. An alternative, more chemical approach, is to precipitate out colloidal quantum dots (which are protected by a layer of organic surfactant molecules, as described above for colloidal metals) as 3-D crystalline superlattices. Such a quantum dot superlattice has been generated for size-selected colloidal CdSe nanoclusters.

8.3.3 Photoluminescence in Semiconductor Clusters

The optical properties of silicon and other semiconductor clusters have been found to change dramatically as a function of the size of the cluster. As the cluster size is reduced, the electronic excitations shift to higher energy and the oscillator strength becomes concentrated into a small number of transitions. These phenomena arise from quantum confinement of the electrons in the finite particles and are due to size-induced changes in the electronic density of states, as discussed above.

Study of the photoluminescence of silicon clusters has shown that, for cluster radii (R) in the size range $R = 2$–100 nm, continuous band structure evolves from discrete, molecule-like states (i.e. the wavefunctions are size dependent). Small semiconductor clusters have excited electronic states with long lifetimes (≈ 100 ns) due to the relatively large HOMO–LUMO gaps that they generally possess. This is in contrast to metal clusters: for example Cu_3 is the only metal cluster which is known to have an excited state lifetime of more than 1 ns. The radiative lifetime decreases as R increases. The radiative transition is an $ns \rightarrow np$ transition for direct band gap semiconductors, such as GaAs and CdS.

For irradiation above the band gap, if $R \gg \lambda$, scattering dominates over absorption (this is a cluster analogue of the fact that reflection is favoured over absorption on flat bulk surfaces. If $R \ll \lambda$, however, electric dipole allowed absorption dominates (which is analogous to molecular spectroscopy) and discrete structures are expected in the 1–10 nm range. Finally, if $R \approx \lambda$, size-dependent resonances occur due to matching of the particle size with the wavelengths of specific electromagnetic modes.

Considering a particle in a semiconducting sphere, the lowest energy excited state is given as a product of the 1s electron and hole wavefunctions and, assuming no electron-hole correlation, is given by:

$$E(R) = \frac{\hbar \pi^2}{2R^2}\left[\frac{1}{m_e} + \frac{1}{m_h}\right] - \frac{1.8e^2}{\varepsilon_1 R} + \text{polarization terms} \qquad (8.1)$$

where R is the sphere radius, ε_1 is the dielectric constant of the sphere, m_e is the effective electron mass and m_h is the effective hole mass. The term in $1/R^2$ is the (quantum) kinetic localization energy and the term in $1/R$ is the electrostatic potential energy. The $1/R^2$ term dominates at small R, so the energy level increases in energy as R becomes small—i.e. there is a blue shift. In the photoluminescence. This causes a change in the luminescence spectrum of Si from the infrared region, for bulk silicon, to the visible region for silicon clusters.

As chemistry (i.e. reactivity) is critically influenced by electronic state, and the electronic structures of semiconductor particles vary with size, one would expect the reactivity of semiconductor clusters also to be size-dependent. The experiments of Smalley, Jarrold and others, on silicon and other semiconductors, confirm this conclusion (see Chapter 6).

8.3.4 Gas Sensors

Large metal oxide nanoclusters (or nanocrystals), derived from ZnO, SnO_2 and other compounds have also found use as gas sensors, since the absorption of small amounts of gas molecules are known to affect their conduction properties by influencing the hole or electron carrier density.

8.4 SUPPORTED CLUSTERS—DEPOSITION, DIFFUSION AND USES

8.4.1 Supported Clusters

Although it is interesting to study isolated clusters in molecular beams, for any useful applications we will need a lot of clusters and some method of storing and transporting them. In particular, they will have to be stabilized against coalescence (bare metal clusters, for example, will fuse spontaneously to form bulk metal). This can be achieved either by coating them with molecules (ligands) which form a protective sheath (as in colloidal metal particles) or by rendering them immobile by depositing them onto, or into, a support. These supports may be (relatively) inert surfaces (e.g. graphite, silicon, oxides) or the insides of porous materials, such as zeolites (aluminosilicates with tunnels and cavities of varying sizes), aluminium phosphates (ALPOs), clays, or porous carbon. Studies of supported clusters offer the possibility of studying many processes (such as catalysis) on the nanometre scale.

Experiments for studying reactivity and catalysis on supported clusters have already been discussed (Section 5.2.5), as have the methods used for determining their structures (Section 4.6.2). I shall concentrate here on methods for depositing clusters, mechanisms of cluster diffusion and applications of cluster deposition.

8.4.2 Methods for Depositing Clusters

In order to fabricate nanoscale devices for opto-electronic applications, it is necessary to deposit size-selected nanoparticles (i.e. clusters) onto a suitable substrate, with high lateral resolution. Over the past decade, many researchers have studied the deposition of clusters and their subsequent (post deposition) structures and diffusion properties. The easiest way to size-select clusters is to use magnetic (or electrostatic) deflection of a beam of charged clusters. Such charged clusters can then be focused onto a relatively small area of the substrate. The impact energy of the size-selected clusters on the substrate can then be adjusted by varying the accelerating potential between the cluster source and the substrate.

8.4.2.1 Low Energy Deposition

If one wishes to deposit clusters onto a substrate without the clusters themselves breaking up or the surface morphology of the substrate being disrupted, the clusters must be deposited with a low impact energy. This is particularly important for small clusters, where the cluster has fewer atoms (and therefore fewer vibrational energy modes) to absorb the impact energy without fragmenting. In theory, this could be achieved by using very low accelerating potentials, but this would lead to much more diffuse beams with much reduced spatial resolution.

On the basis of Molecular Dynamics (MD) simulations, Landmann proposed (in 1993) that by cooling the substrate and adsorbing a thin film of a rare gas, the energy of an impacting cluster could be dissipated. In this 'soft landing' approach, the rare gas acts as a braking layer, as well as aiding the post-impact cooling process, via evaporation. Landmann's predictions have been verified experimentally by Harbich and co-workers who used an Ar braking layer to soft land Ag_7 clusters. Subsequent STM investigation of the surface showed the Ag_7 clusters to be intact.

8.4.2.2 High Energy Deposition

There have been a number of experimental and theoretical studies of the effect of high energy impact of a cluster on a substrate. The effects observed depend on the impact energy (which depends on the accelerating potential applied to the charged cluster beam) and the nature of the materials comprising the cluster and the substrate.

On the basis of MD simulations of the impact of metal clusters on metal substrates, Hsieh *et al.* found that for moderate impact energies (10s–100s of eV) the clusters flatten out on impact and cause little surface damage, though there may be some surface mixing between atoms of the cluster and the substrate. For a high energy (keV) impact, a crater is generally formed on the surface, which, due to the high local temperatures caused by the impact, often anneals to form a crater hollow which is filled by cluster material. However, high energy (1 keV) impact of 'soft' metal clusters (e.g. Al_N^+) on a hard metal (e.g. Ni) substrate merely leads to 'splatting' (flattening and lateral spread) of the cluster and little disruption of the Ni surface. Conversely, at the same impact energy, a hard Ni cluster impacting on a soft Al substrate remains intact and is embedded within the substrate. The high local temperature generated by the impact causes the Al substrate to anneal and close-up over the embedded cluster.

Betz and Husinsky have shown (again on the basis of MD simulations) that in collisions between metal clusters and metal substrates there is a rapid initial rise in temperature (a *thermal spike*) at the impact site, with the maximum temperature of this spike depending on the energy *per atom* of the incident cluster. The cluster (and substrate) cool rapidly, with the cooling rate being strongly size-dependent: large clusters cool more slowly. Betz and Husinsky proposed that, by carefully tailoring the temperature and duration of the thermal spike, clusters could be used for epitaxial film growth or surface sputtering—and clusters have indeed been used for these purposes.

For covalent substrates and clusters, due to the strong, directional bonds, there is little annealing following impact. Smith and co-workers have studied the impact of C_{60} on a Si surface (at 5 keV) and have found that there is complete fragmentation of the cluster and that the crater (which is not annealed) is lined with amorphous material and has a raised rim. Collisions with layered covalent substrates, such as graphite, result in a large elastic compression and the generation of transverse lattice waves or ripples. Carbon atoms displaced by the impact often become lodged as interstitial atoms between the graphite planes, resulting in bumps in the surface layer, which can be detected by microscopy techniques.

Finally, considerable excitement was generated in the late 1980s by research at Brookhaven National Laboratory in the USA, where heavy water clusters $(D_2O)_N$ ($N \approx 1000$) were collided at exteremely high (MeV) impact energies onto deuterated Ti substrates and high D–D fusion rates were reported. However, subsequent studies by other groups have failed to reproduce these intriguing results.

8.4.2.3 Chemical Deposition

As discussed in Section 8.3, colloidal clusters can be grown chemically in water or organic solvents. These clusters can be deposited on the surface of a substrate simply by dropping some of the colloidal solution onto the substrate and allowing the solvent to evaporate. Chemical deposition methods can be used for a variety of metal and semiconductor clusters (if soluble colloidal solutions can be generated). The main advantages of chemical deposition are that large amounts of cluster materials can be deposited in a short time and that under equilibrium conditions, long-range ordered 2-D and 3-D arrays of clusters can be deposited. Disadvantages of this method include decreased cluster size-selectivity and possible co-deposition of solvent and impurity molecules along with the clusters.

8.4.3 Cluster Diffusion

When clusters are deposited at low impact energies on substrates that are at room temperature, they are often able to diffuse across the surface of the substrate. Simple models of thin film growth by atom deposition have traditionally assumed that clusters with more than a few atoms are unable to diffuse, but this assumption has been shown to be wrong by many studies. For example, studies of antimony cluster deposition indicate cluster diffusion rates as high as 10^{-8} cm^2 s^{-1} for clusters as large as Sb_{4000}!

8.4.3.1 Models of Cluster Diffusion

In earlier studies of the diffusion of Rh_N clusters ($N \leq 12$) on the tip of a *field ion microscope*, Kellogg found that cluster diffusivity increased with increasing cluster size, with notable oscillations. These findings were explained in terms of a *Periphery Diffusion* (PD) mechanism, wherein atoms on the periphery of the cluster diffuse, causing a change in the cluster's centre of mass and, hence, cluster diffusion. Wang found similar behaviour for the diffusion of iridium clusters Ir_N ($N \leq 39$), though the most compact (Ir_7 and Ir_{19}) clusters were found to diffuse 1,000 times faster than the non-compact clusters. This enhanced diffusion rate was attributed to a *gliding* mechanism of the cluster as a whole. The relatively high diffusion rates measured for large silver clusters (Ag_{100}–Ag_{720}) on Ag(100) were explained by Wang in terms of an *Evaporation–Condensation* (EC) model (i.e. the cluster diffuses by the repeated evaporation and condensation of atoms), rather than to the PD or gliding mechanisms.

For large 3-D clusters, both the PD and the EC mechanisms predict diffusion rates which are lower than those measured experimentally. Bardotti and colleagues proposed that, for such clusters, diffusion involves a collective motion of the atoms in the cluster (as in Wang's gliding mechanism): with the cluster effectively rolling over the surface. On the basis of MD simulations, the same group have proposed a mechanism for cluster diffusion that is driven by lattice vibrations (i.e. phonons) in both the substrate and the cluster, resulting in Brownian motion of the cluster on the substrate. In this model, as well as depending on the temperature of the cluster and substrate, the cluster diffusion rate is dependent on the lattice mis-match between the cluster and the substrate, since a good lattice match (which may be described as 'commensurate' or 'epitaxial') leads to stronger cluster–substrate binding and lower diffusivity. These predictions have been confirmed by the experiments of Palmer and co-workers who found that small silver clusters have low diffusivity on graphite, because of an accidental matching of latting spacings in the cluster and the substrate. In larger silver clusters, however, the cluster Ag–Ag bond lengths are longer than for small clusters, so the cluster–substrate lattice matching is not so good and the clusters diffuse more rapidly.

Finally, Luedke and Landmann have studied (theoretically) the diffusion of deposited thiol-passivated colloidal gold clusters. High diffusion constants were attributed to cooperative motions of the alkylthiol ligands, which has a 'lubricating' effect, thereby reducing the energy barrier to diffusion.

8.4.3.2 Cluster Aggregation

The tendency of deposited clusters to coalesce to a certain extent removes the benefit of size-collecting clusters prior to deposition. Cluster coalescence is governed by the shape and size of the cluster, as well as the material of which they are composed. In order to coalesce, the component atoms (or molecules) must overcome an energy barrier in order to diffuse around their original cluster and join the other. Films of strongly covalently-bonded clusters, such as fullerenes, are stable to coalescence, since there are large energy barriers, associated with breaking C–C bonds, to overcome. Small metal clusters, however, coalesce rapidly—with some calculations indicating that there is no barrier to coalescence for non-passivated clusters. Large metal clusters also eventually coalesce, though coalescence may be slow. Goldby and co-workers found that Ag_{50-400} clusters coalesced rapidly on graphite (forming 14 nm diameter islands, which slowly scinter to form larger particles). In the case of antimony clusters, however, Perez demonstrated that, while Sb_{250} clusters rapidly coalesced to form 3–5 nm particles, larger Sb_{2300} particles did not coalesce on the timescale of their experiment.

Control of the level of aggregation/coalescence can be achieved in a number of ways: cooling the substrate to lower cluster diffusion; pinning clusters to the substrate by using higher deposition energies; chemically tethering them to the surface; and coating the clusters with passivating ligands (see Section 8.2).

Monte Carlo models have been developed to simulate the diffusion and aggregation of surface-deposited clusters and the eventual formation of thin films (see Section 8.4.5.1). The *Deposition–Diffusion–Aggregation* (DDA) model of Jensen *et al.* was introduced to study island growth on surfaces. In the DDA model, clusters are deposited at random onto the surface. The clusters then diffuse (as rigid bodies), via a random walk, until they meet and coalesce with another species. In the simplest model, such coalescence leads to the nucleation of an island, which is treated as being fixed. Islands form and grow from these nuclei. An extension of the model allows for size-dependent mobility of the islands, which generally results in a bimodal distribution of island sizes. If the strength of the bonding between the deposited clusters is large, there is no edge diffusion (movement of clusters around the growing island) and the clusters stick together in the way that they first collide—forming fractal or dendritic islands. For lower inter-cluster interactions (or higher cluster/substrate temperatures), more edge-diffusion takes place, resulting in more compact clusters.

8.4.4 Study of Silver Cluster Deposition on Graphite

Palmer, Smith and colleagues have carried out a detailed combined experimental and theoretical study of the deposition and diffusion of charged silver clusters Ag_N^+ ($N = 3$–400) on a graphite substrate, at various impact energies.

At low impact energies (up to a few eV per cluster atom—e.g. Ag_{400} at an impact energy of 500 eV) the clusters are not greatly deformed on landing and no defects are induced in the substrate. The clusters diffuse across the flat terraces of the graphite substrate, coalescing to form islands, which are still mobile up to diameters of around 14 nm (as observed by Goldby *et al.*). The initially deposited clusters and mobile islands are trapped at steps on the graphite surface, though there is evidence for limited diffusion of clusters along the steps. Increasing the deposition flux leads to an increase in the average number of islands and a corresponding decrease in average size.

For medium energy impacts (10s of eV per atom), the clusters flatten and there is some surface damage. Above a threshold energy of 10.4 eV/atom (e.g. an impact energy of approximately 500 eV for Ag_{50}) the cluster penetrates the graphite surface and, hence, becomes pinned to the surface, thereby inhibiting post-impact diffusion. Palmer and co-workers have calculated that this threshold impact energy (per atom) results in approximately 4 eV of energy being imparted to a surface carbon atom, which is sufficient to break one or two C–C bonds, with the cluster being pinned by bonding to the damaged surface.

Figure 8.8 The result of a simulation of high energy impact of a silver cluster with graphite. The cluster lies in a crater in the graphite substrate.

For high impact energies (100s of eV per atom), the clusters are implanted into the graphite substrate, with extensive surface and sub-surface damage of the substrate, restructuring of the cluster and transverse elastic waves being excited in the graphite lattice. MD simulations indicate that, at an impact energy (E_i) of 300 eV, an Ag_3 cluster is implanted as far as the third to fifth graphite layer and that a bump appears on the graphite surface, either due to the formation of interstitial carbon atoms or to interlayer C–C bonding. At somewhat lower energies ($E_i \approx$ 150 eV) the extent of implantation depends on the orientation of the cluster and the nature of the impact site. For larger clusters (Ag_{20-200}), high energy impacts result in the cluster lying at the bottom of an open crater (see Figure 8.8), the walls of which are composed of amorphous carbon. For these clusters, Palmer and colleagues have found that the implantation depth (D_i) of the cluster depends on both the impact energy and the cross sectional area of the cluster, which (assuming a pseudo-spherical cluster geometry) gives the following relationship:

$$D_i \propto \frac{E_i}{N^{2/3}} \qquad\qquad (8.2)$$

8.4.5 Applications of Deposited Clusters

In this section, I will discuss the applications of cluster deposition to the fabrication of thin films, surface circuits and surface lithography. Some other applications of surface-supported clusters (e.g. in catalysis, optics and electronics) have been discussed previously.

8.4.5.1 Growing Thin Films by Cluster Deposition

Perez and co-workers have used soft landing techniques to deposit clusters and generate thin films: as more clusters are deposited, the clusters join up and a rough, thin film is produced. Perez has found that the physical properties of these films (e.g. hardness, electrical conductivity, optical absorption) sometimes depend on the sizes of clusters deposited—in other words the thin film displays a *cluster memory effect*. Such an effect is to be expected for films of fullerenes, for example, where the clusters remain distinct and identifiable upon film formation. For films generated by deposition of transition metal clusters however, such memory effects are at least partially washed out by cluster coalescence. The cluster origin of such transition metal films can, however, be seen in the fact that they are more porous than films generated by conventional single atom deposition.

In contrast to Perez's soft landing approach, Haberland has used high energy (up to 30 keV) deposition of large metal clusters M_N (M = Ag, Al, Co, Cu, Fe, Mg, Mo and Ti; N = 500–10,000) to grow thin films on a variety of substrates, such as metals, ceramics, and polymers. The same techniques have been applied to grow films of Si and TiN. As mentioned above, for metal clusters, the high impact energy leads to lateral spreading and thermal annealing. This results in high quality epitaxial films which are smooth and have low porosity, even when generated at low substrate temperatures. Due to collision-induced cluster–surface mixing at the interface, the films adhere well to the substrate—even to PTFE (Teflon®) and paper!

In another development of great technological importance, Haberland has shown that a fine beam of energetic clusters can be directed onto a film or wire in order to repair any defects—we could describe this as 'nanowelding'. Thus, copper clusters can be used to fill micron sized holes which sometimes form during the production of microelectronic integrated circuits. The advantage of Haberland's repair method is that it can be carried out at moderately low temperatures (80°C), thereby avoiding damage of the silicon chip.

The fact that cluster impact produces localized surface damage and heating has been exploited by Yamada *et al.* to clean and polish surfaces by bombarding them with clusters of atoms or molecules which are volatile (e.g. Ar and CO_2 clusters) so that the substrate surface is not contaminated by atoms/molecules from the clusters themselves.

8.4.5.2 Nanostructure Fabrication

The high diffusivity of soft-landed clusters on substrates, such as graphite, presents opportunities for depositing clusters and subsequently organizing them on the surface (either using their inherent thermal diffusion or 'manually' moving them using an STM or AFM tip) to fabricate structures on the nanoscale. Palmer and colleagues have noted the migration of silver clusters, deposited at low impact energies, and their pinning at steps on a graphite surface. As shown in Figure 8.9, chains of clusters can be formed at such steps, opening up the possibility of constructing nanowires and other components for nanoscale electronic devices.

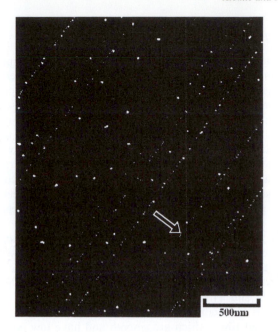

Figure 8.9 SEM micrograph of Ag$_{400}$ clusters deposited at 500 eV onto a stepped graphite surface. (The coverage is 4.7×10^8 (clusters) cm^{-2}, deposited at a rate of 1.3×10^6 cm^{-2} s^{-1}.) Chains of clusters, which are pinned to steps on the graphite surface (one such step is indicated by the arrow), can clearly be seen, as well as larger silver islands which lie on the terraces.

Defects in such nanowires could be repaired using Haberland's nanowelding technique.

This idea can be extended to enable one to build nanoscale arrays of clusters, with desired connectivity, on a pre-patterned substrate—in fact (as mentioned in Section 8.3.2) Heath has used such a technique to obtain a regular array of surface-supported quantum dot nanoparticles. Alternatively, medium impact cluster deposition can be used to pin clusters to a substrate, which can then act as nucleation sites for island growth at selected sites.

8.4.5.3 Use of Clusters as Masks for Surface Etching

Palmer and co-workers in Birmingham, in collaboration with Tada and Kanayama in Tsukuba (Japan), have employed deposited clusters as nanometre scale tools in the fabrication of semiconducting devices, rather than as the devices themselves. Size-selected silver clusters are deposited (at impact energies sufficiently high to pin them) onto the surface of a silicon wafer, which is subsequently plasma-etched using SF$_6$ gas at low temperature (−130°C). The plasma eats away the Si that is not protected by the silver cluster masks, generating an array of Si pillars with diameters of around 15 nm and heights of 100 nm, as shown in Figure 8.10. These silicon pillar arrays are of great interest for photoluminescent device applications, making use of the size-dependent colour of photoluminescence of Si nanostructures discussed in Section 8.3.3.

It was observed that the diameters of the Si pillars ($\approx$ 15 nm) were significantly larger than the deposited clusters (even if flattened out). Further studies revealed that changing the size of the deposited clusters (even by a factor of twenty) does not change the diameter of the Si pillars, though the efficiency of pillar formation does show some size-dependence. Thus, no clusters are formed using Ag_{200} clusters, a small number for Ag_{300}, and using Ag_{720} (or larger) clusters, nearly every cluster deposited results in a pillar after etching.

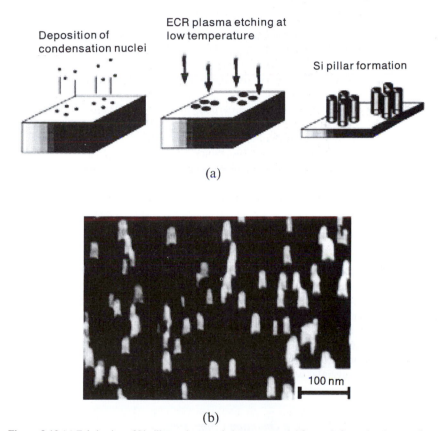

(a)

(b)

Figure 8.10 (a) Fabrication of Si pillars using metal clusters as nuclei for mask formation (see text for details). (b) SEM image of Si pillars.

In light of the above observations, it is believed that the silver clusters do not themselves act as etch masks, but rather that they are nucleation sites for the condensation of etch products (Si_xF_y and SiF_z), which create a self-forming etch mask, having a greater diameter than the underlying cluster. The size-dependence of pillar formation probability could be explained in terms of either a size-dependent nucleation efficiency of the Ag clusters and/or a cluster size which is sufficient for it not to be sputtered away (under the plasma etching conditions) before the etch mask can form. Interestingly, using gold clusters of a similar size to the Ag clusters as etch masks, the pillar diameter is reduced to around 8 nm, indicating a chemical dependence of pillar width.

8.4.5.4 Silver Clusters in Photography

The black-and-white photographic process, which has been around for over a century, involves the local photodissociation of light-sensitive silver halides (e.g. AgBr), creating neutral silver and halogen atoms (radicals):

$$Ag^+Br^- + h\nu \rightarrow Ag^\bullet + Br^\bullet$$

This process is repeated and some of the silver atoms cluster together to form a so-called latent-image speck, which accelerates or catalyses the reduction (in the photographic development process) of exposed AgBr microcrystals to silver metal—a process which is slow in the absence of such latent-speck initiation. By studying the subsequent development of photographic films, on which size-selected silver clusters have been bombarded, Fayet and colleagues established (in 1985) that clusters of at least four atoms (Ag_4^+) are required to initiate silver bromide reduction—i.e. to allow the plate to be developed.

It has long been known that the silver latent-image specks on an exposed film, if not rapidly developed, can undergo oxidation, leading to deterioration of the latent image. This has been explained in terms of the lower ionization energies of small silver clusters, compared with the bulk metal.

8.4.5.5 Nanowires in Zeolites

Edwards and Anderson have managed to synthesize a number of highly reactive charged alkali metal clusters inside zeolite cavities and have demonstrated how, as more and more metal atoms are added, the zeolitic powder turns blue and finally black and that the zeolites become semiconducting as electrons can hop easily from cluster to cluster. The ultimate goal of this work is to form chains of metal atoms in the zeolites and thus assemble dense bundles of one-dimensional conductors or nanowires. For application purposes, however, less reactive metals would have to be used, so there is now considerable interst in loading zeolites and other porous materials with noble metal atoms and clusters (as discussed in Section 8.2).

8.4.5.6 Deposited Clusters in Quantum Computing?

Smith and colleagues have suggested that fullerenes containing odd electron species, as in $N@C_{60}$ or $P@C_{60}$ could be deposited in arrays on a silicon substrate and used as bits (so-called 'q-bits') for quantum computing, making use of coupling between electron spins in adjacent endohedral fullerene complexes.

8.5 FURTHER READING

Alivisatos, A.P., 1996, Semiconductor clusters, nanocrystals, and quantum dots. *Science*, **271**, pp. 933–937.

Attfield, J.P., Johnston, R.L., Kroto, H.W. and Prassides, K., 1999, New science from new materials. In *The Age of the Molecule*, edited by Hall, N., (London, Royal Society of Chemistry), pp. 181–208.

Broglia, R.A., 1994, The colour of metal clusters and of atomic nuclei. *Contemporary Physics*, **35**, pp. 95–104.

Edwards, P.P., Johnston, R.L. and Rao, C.N.R., 1999, On the size-induced metal-insulator transition in clusters and small particles. In *Metal Clusters in Chemistry*, Vol.3, edited by Braunstein, P., Oro, L.A. and Raithby, P.R., (Weinheim, Wiley-VCH), pp. 1454–1481.

Kanemitsu, Y., 1995, Light emission from porous silicon and related materials. *Physics Reports*, **263**, pp. 2–91.

Kiely, C.J., Fink, J., Brust, M., Bethell, D. and Schiffrin, D.J., 1998, Spontaneous ordering of bimodal assemblies of nanoscopic gold clusters. *Nature*, **396**, pp. 444–446.

Kreibig, U. and Quinten, M., 1994, Electromagnetic excitations of large clusters. In *Clusters of Atoms and Molecules*, Vol.II, edited by Haberland, H., (Berlin, Springer-Verlag), pp. 321–359.

Palmer, R. E. 1997, Welcome to clusterworld. *New Scientist* (22 February), pp. 38–41.

Rao, C.N.R., Kulkarni, G.U., Thomas, P.J. and Edwards, P.P., 2000, Metal nanoparticles and their assemblies. *Chemical Society Reviews*, **29**, pp. 27–35.

Schmid, G., 1999, Nanosized clusters on and in supports – perspectives for future catalysis. In *Metal Clusters in Chemistry*, Vol. 3, edited by Braunstein, P., Oro, L.A. and Raithby, P.R., (Weinheim, Wiley-VCH), pp. 1325–1341.

Simon, U., 1999, On the possibility of single electronics based on ligand-stabilized metal clusters. In *Metal Clusters in Chemistry*, Vol.3, edited by Braunstein, P., Oro, L.A. and Raithby, P.R., (Weinheim, Wiley-VCH), pp. 1342–1363.

Sugano, S., 1991, *Microcluster Physics*, (Berlin, Springer-Verlag), pp. 104–117.

Whetten, R.L., Shafigullin, M.N., Khoury, J.T., Schaaff, T.G., Vezmar, I., Alvarez, M.M. and Wilkinson, A., 1999, Crystal structures of molecular gold nanocrystal arrays. *Accounts of Chemical Research*, **32**, pp. 397–406.

APPENDIX

Answers to Exercises

In this appendix, brief answers are given to the questions posed at the end of each chapter. In most cases, the reader is referred to specific sections for more details.

CHAPTER 1

1.1 From Equation (1.6) (Section 1.2.1), the fraction of surface atoms:

$$F_s = \frac{N_s}{N} = 4N^{-1/3} \tag{A1.1}$$

drops below 0.1 (i.e. 10%) when $4N^{-1/3} < 0.1$, corresponding to N > $(0.1/4)^{-3}$ (i.e. N > 64,000 atoms).

1.2 The smaller the packing fraction (f) the greater the number of surface atoms (N_s) for a given total number of atoms (N) and, hence, the higher the fraction of surface atoms (F_s). Equation (1.1) (Section 1.2.1), becomes:

$$V_c = NV_a^* = \frac{NV_a}{f} \tag{A1.2}$$

where V_a^* is the effective atomic volume and V_a is the volume of the spherical atom. In terms of the radii of the spherical cluster and atom:

$$N = \frac{V_c}{V_a^*} = \frac{fV_c}{V_a} = \left(\frac{R_c}{R_a}\right)^3 f \tag{A1.3}$$

which is equivalent to Equation (4.31) and enables us to write:

$$R_c = \left(\frac{N}{f}\right)^{1/3} R_a \tag{A1.4}$$

If we retain the other assumptions of the Spherical Cluster Approximation, Equation (1.5) becomes:

$$N_s = \frac{4\pi\left(N/f\right)^{2/3} R_a^2}{\pi R_a^2} = 4\left(\frac{N}{f}\right)^{2/3} \tag{A1.5}$$

and Equation (1.6) becomes:

$$F_s = \frac{N_s}{N} = 4f^{-2/3}N^{-1/3} \tag{A1.6}$$

Thus, for fcc packing ($f_{fcc} = 0.74$) we get:

$$F_s = 4f^{-2/3}N^{-1/3} \approx 4.89N^{-1/3} \tag{A1.7}$$

1.3 From Section 1.3.5.2, the MS peak intensity (I_N) depends on the following factors:

$$I_N = F(P_N, S_N, X_N, D_N) \tag{A1.8}$$

where P_N is the cluster production efficiency, S_N is the cluster stability; X_N is the ionization efficiency, and D_N is the detection efficiency. Strictly speaking, the cluster stability is not in itself an experimental factor, as it depends on the intrinsic nature of the material from which the cluster is constituted. However, altering experimental conditions (in particular varying the pressure and/or temperature in the cluster generation/condensation region) can alter the distribution of peaks in the mass spectra. (See Section 1.3 for details.)

1.4 The main method for preparing clusters of rare gas elements (which are already in the vapour phase at standard temperatures and pressures) is using a supersonic free-jet nozzle source.

For refractory (high melting point) metals, such as iron, laser-vaporization-flow condensation; pulsed-arc; ion sputtering; and magnetron sputtering sources may be used. (See Section 1.3.1 for details.)

CHAPTER 2

2.1 The equilibrium bond length and dissociation energies of $(Ar_2)^*$ and Ar_2^+ are very similar because electronically excited rare gas dimers are effectively Rydberg states of the dimer (see Section 2.4.1), consisting of a loosely bound electron and a covalently bonded Ar_2^+ core, as depicted in Figure 2.9.

2.2 The zero-point vibrational energy of a diatomic molecule (modelled by a harmonic oscillator) is given by:

$$\text{ZPE} = \frac{1}{2} h\nu \tag{A2.1}$$

where the vibrational frequency:

$$\nu = \sqrt{\frac{k}{\mu}} \tag{A2.2}$$

k is the stretching force constant and μ is the reduced mass $(m_1 m_2)/(m_1+m_2)$, where m_1 and m_2 are the masses of the two atoms. In atomic mass units (the actual units used do not matter, as they will cancel out): $\mu(^4\text{He}_2) = (16/8) = 2$; $\mu(^3\text{He}^4\text{He}) = (12/7) = 1.71$ (2 d.p.) and $\mu(^3\text{He}_2) = (9/6) = 1.5$. If we assume that k is independent of the mass and we define $\text{ZPE}(^4\text{He}_2) = 1$, then we obtain: $\text{ZPE}(^3\text{He}^4\text{He}) = \text{ZPE}(^4\text{He}_2) \times [\mu(^4\text{He}_2)/\mu(^3\text{He}^4\text{He})]^{1/2} = 1 \times (2/1.7143)^{1/2} = 1.08$; and $\text{ZPE}(^3\text{He}_2) = (2/1.5)^{1/2} = 1.15$. Thus, the lower the isotope masses, the higher the zero-point vibrational energy.

CHAPTER 3

3.1   Table 3.1 (see Section 3.1.3) shows that small non-polar molecules (such as N_2) have low intermolecular binding energies, while larger molecules (e.g. C_6H_6) have significantly higher interactions—which shows up in the respective potential well depths for the dimers (92 K for $(N_2)_2$ and 440 K for $(C_6H_6)_2$). The boiling points of the bulk materials (77 K for N_2 and 353 K for C_6H_6) follow the same trend as the well depth. While the bonding in clusters of N_2, C_6H_6 and H_2O all have contributions from dispersion forces, nitrogen and benzene clusters have important quadrupolar interactions, and C–H$\cdots\pi$ bonding may contribute to the stabilization of benzene clusters. A major contribution to the bonding in water clusters is hydrogen bonding, which gives rise to significantly higher well depths—the effective pair potential well depth for $H_2O\cdots H_2O$ is 2400 K, which is also consistent with the relatively high boiling temperature (373 K) of liquid water.

3.2   Although the electron diffraction patterns for medium-sized water clusters (with N in the range 1,500–2,000) are similar to those for the metastable, low pressure cubic phase of bulk ice, smaller clusters (with $N \lesssim 300$) often have amorphous, or highly disordered structures, consisting of three-, to six-membered hydrogen-bonded rings (see Section 3.3.4). These structures, which are reminiscent of those found in amorphous ice, are consistent with conclusions based on the IR spectra of small water clusters (with $N \approx 100$), as well as theoretical calculations. Structures based on the dodecahedral $(H_2O)_{20}$ unit have been proposed, though an alternative fused pentagonal prismatic structure has been proposed for the 20-molecule cluster itself.

   The dominance of surface effects for small water clusters means that structures based on the bulk ice structure (consisting of exclusively 6-membered hydrogen-bonded rings) will have a large number of dangling hydrogen bonds. By adopting non-ice-like structures, with a certain number of small hydrogen-bonded rings, there will be more hydrogen bonding, a lower surface energy and, hence, a lower cluster energy. As the clusters get larger, there is a competition between the preference for bulk-like structure in the centre of the cluster and reducing the number of dangling hydrogen bonds at the surface. Calculations have shown that the $(H_2O)_{1000}$ cluster, for example, has an ordered (ice-like) interior, a disordered surface, and strained ice-like structure in the sub-surface region.

CHAPTER 4

4.1   The Liquid Drop Model (LDM, see Section 4.2) is the simplest model for metal clusters. It is a classical electrostatic model in which a metal cluster is approximated by a uniform conducting sphere. Examples of liquid drop behaviour can be found in the ionization energies (IP) and electron affinities (EA) of large metal clusters, which are given by:

$$IP(R) = W + \left(\frac{3}{8} \times \frac{1}{4\pi\varepsilon_0 R} \right) \tag{A4.1}$$

$$EA(R) = W - \left(\frac{5}{8} \times \frac{1}{4\pi\varepsilon_0 R} \right) \tag{A4.2}$$

In general, liquid drop behaviour is manifested as a linear dependence of a cluster property on the inverse radius (or, equivalently on $N^{-1/3}$). Other properties of metal clusters that show this behaviour include HOMO-LUMO gaps and surface plasmon frequencies.

For small clusters, there are significant deviations from the predicted $1/R$ ($N^{-1/3}$) dependence of ionization energies, electron affinities, and HOMO–LUMO gaps. Thus, the LDM cannot reproduce the even–odd alternation observed in the IPs of small sodium clusters. It also fails to explain the occurrence of magic numbers observed in the mass spectra of alkali metal clusters. These deviations are due to quantum size effects (i.e. electronic shell filling).

For many transition metals (such as iron and nickel), where ionizations involve the removal of quite tightly bound d electrons, only a small variation of the IP and EA are observed wich changing cluster nuclearity. In these clusters, magnetic effects, due to electronic spin-spin interactions are also important.

Another deviation from the LDM is shown by the ionization energies of mercury clusters, when plotted against $1/R$, rather than following a simple LDM straight line, exhibit a clear discontinuity, which has been attributed to a non-metal $\rightarrow$ metal transition.

4.2 The jellium model predicts magic numbers based on complete filling of electronic energy levels. In this question, one must determine which sizes of cluster (*i.e.* number of atoms, N) are consistent with the closed shell electron counts. These N values are the magic numbers and should correspond to clusters of enhanced stability (relative to clusters with (N–1) or (N+1) atoms).

The ordering of energy levels is: $1s < 1p < 1d < 2s < 1f < 2p < 1g$. As for atomic orbitals, the degeneracy of the jellium orbitals with angular momentum quantum number ℓ is $(2\ell+1)$ and the number of electrons in each jellium shell is $(4\ell+2)$. Jellium shell closings will therefore be seen for $N_e = $ 2, 8 (2+6), 18, 20, 34, 40, 58 ... electrons.

(i) For uncharged alkali metal clusters (M_N), since each metal has one valence s electron, $N_e = N$. In the range $N = 2$–50, the magic numbers will be: $N = 2, 8, 18, 20, 34$ and 40 (*e.g.* Li_8, Na_{20} ...).

(ii) Cu has the configuration $[Ar]3d^{10}4s^1$ but only the valence s electron is extensively delocalized in Cu clusters, so as far as the jellium model is concerned, Cu is a one-electron metal. For Cu_N^+, $N_e = N$–1, so the magic numbers correspond to: $N = 3, 9, 19, 21, 35$ and 41 (*e.g.* Cu_{19}^+).

(iii) Al has the configuration $[Ne]3s^23p^1$. Including all 3 valence electrons in the jellium model, $N_e = 3N+1$, so the only magic number in this size range is at $N = 13$ ($N_e = 40$), *i.e.* Al_{13}^-.

4.3 According to the jellium model (see Section 4.3), the ordering of jellium energy levels is: $1s < 1p < 1d < 2s < 1f < 2p < 1g$. Na_{19}^+ and Na_{21}^+ have 18 and 20 valence electrons, respectively, both corresponding to closed shell jellium configurations ($Na_{19}^+ = (1s)^2(1p)^6(1d)^{10}$; $Na_{21}^+ = (1s)^2(1p)^6(1d)^{10}$ $(2s)^2$) and therefore have pseudo-spherical structures.

Na_{17}^+ (16e) and Na_{23}^+ (22e) have open jellium shells and lower their energy by ellipsoidal distortion (see Section 4.3.3). Na_{17}^+, with the configuration $(1s)^2(1p)^6(1d)^8$, is an oblate spheroid, raising the energy of the single empty d orbital, while Na_{23}^+, with the configuration $(1s)^2(1p)^6$ $(1d)^{10}(2s)^2(1f)^2$ is a prolate spheroid, lowering the energy of the single filled f orbital.

4.4 The mass spectra (MS) of small sodium clusters (from 2 to hundreds of atoms—as in the experiments of Knight *et al.*) show discrete peaks ('magic numbers') due to the complete filling of jellium shells (see Section 4.3).

In the region N = hundreds to around 2,000 atoms, Martin measured a mass spectrum (see Section 4.4) consisting of periodic oscillations in intensity which are approximately evenly spaced when plotted against $N^{1/3}$. The individual jellium peaks are no longer seen, because for these medium sized clusters the jellium levels are closely bunched together, forming quasi-degenerate bands with gaps inbetween. It is the filling of these bands or 'electronic shells' that give rise to the oscillations. In Martin's experiments the magic numbers correspond to dips in intensity as near-threshold ionization is used.

For N between 2,000 and 25,000 atoms, Martin observed another series of periodic oscillations, again approximately evenly spaced when plotted against $N^{1/3}$, but with greater spacing. These have been explained in terms of filling geometric shells—i.e. complete polyhedral shells of atoms (see Section 4.5). For these large clusters, electronic shell effects are not seen because the bands of jellium levels merge to form a continuum (as in bulk metals) so there is no enhanced stability conferred by one electron count over another. Instead, cluster stabilities (and hence magic numbers) depend on the lowering of surface energy due to filling of geometric shells of atoms. In near threshold ionization mass spectroscopy experiments, these magic numbers again show up as dips in intensity. Analysis of the fine structure of the mass spectra indicates that these large magic number clusters are icosahedral.

CHAPTER 5

5.1 Small molecules can be used as chemical probes of cluster structure, by studying the extent of adsorption as a function of pressure and cluster size (see, for example the ammonia uptake experiments discussed in Section 5.2.3). Plateaux in plots of average uptake vs. pressure (at low pressures) gives information as to the number of sites where adsorption is facile, while higher pressure measurements tells us about how many surface sites are available in total.

5.2 From Equation (5.5), metallic conduction is predicted when $(E_F/N) \leq kT$. For Na_{1000} (using $E_F(Na) = 3.24$ eV), metallic conduction should be observed above a temperature of $T = (E_F/ k.N) = (3.24 \times 1.602 \times 10^{-19}$ J) / $(1000 \times 1.381 \times 10^{-19}$ J $K^{-1}) = 37.6$ K.

5.3 The main evidence for a transition from non-metallic to metallic bonding in mercury clusters (possibly proceeding via semiconducting clusters) is the variation of cluster ionization energies with size (see Section 5.3.2.3). Small clusters have IPs which extrapolate to a bulk ionization energy significantly higher than the work function of Hg metal. In the region of $N = 13–70$ atoms, there is a rapid drop in IP, until the curve hits the prediction of the LDM for metallic clusters.

Other experiments on mercury clusters, such as studies of the size-dependence of the $5d–6p$ autoionization spectrum and the appearance of the surface plasmon mode, provide supporting evidence for an insulator to metal transition beginning at around 13–20 atoms. Photoelectron spectroscopic measurements for mass selected Hg_N^- clusters ($N = 3–250$), however, show that the HOMO–LUMO gap (for the neutral cluster) decreases gradually from ≈ 3.5 eV for Hg_3 to ≈ 0.2 eV for Hg_{250}, with band-gap closure predicted to occur at around $N = 400$.

5.4 In iron and nickel clusters (see Section 5.4.3), the magnetic moments are dynamically decoupled from the lattice, though they are dynamically coupled together via exchange interactions. Thus, the magnetic moments rotate in unison relative to the lattice, under the influence of the applied field, and the cluster behaves as a paramagnet with a large net magnetic moment—i.e. a 'superparamagnet'. At temperatures above a certain, critical temperature (T_c, the counterpart of the Curie temperature in bulk ferromagnets), thermal fluctuations are sufficiently large to overcome the coupling of spins with the applied field and there is no superparamagnetic moment.

CHAPTER 6

6.1   The relative stabilities of fullerene isomers are governed by two factors (see Section 6.2.3.5): the energies of the $N/2$ occupied π-MOs; and the strain energy of the sp^2 framework. The total radial orbital energy is related to the open or closed shell nature of the fullerene, since properly closed shell clusters have large HOMO–LUMO gaps (Δ) and, therefore, the sum of the one-electron π-electron energies will be lower. While small Δ values, or open shells, are usually associated with low stability, however, a large Δ does not always imply greater stability, due to the greater importance of the steric preferences of the sp^2 framework. Another way of considering the electronic stabilization of fullerene isomers is to consider them as superalkenes (see Section 6.2.3.8). If fullerene isomers are drawn such that there are no double bonds in the pentagonal rings and each atom has a valency of four, then the lowest energy isomers tend to be those with the greatest number of benzenoid rings.

Considering the strain energy of the sp^2 backbone, the main structural feature governing isomer stability is the *isolated pentagon rule*, which states that for a given nuclearity the relative energies of fullerene isomers increase with increasing numbers of adjacent pentagons. The buckminsterfullerene isomer of C_{60} is the smallest fullerene with isolated pentagons. Given a choice of isomers with the same number of adjacent pentagons, Raghavachari has said that the most stable isomers will have the most even distribution of hexagon environments.

6.2   The ^{13}C NMR spectra of fullerenes in solution can be used to distinguish between different isomers by consideration of the symmetries of the alternative geometries and how many symmetry-distinct carbon atoms they possess—this tells us how many distinct ^{13}C resonances should be observed—which is then compared with the experimental NMR spectrum. The application of this type of approach to C_{70} is described in Section 6.2.3.1.

6.3   As described in Section 6.3.3, Jarrold has measured the mobilities of mass-selected charged silicon clusters, Si_N^+ ($N = 10$–60), using an ion drift tube. The arrival time distributions for $N = 10$–23, showed a single peak, indicating only one isomer for each nuclearity. In the range $N = 24$–35, two components were be resolved, corresponding to two isomers with significantly different topologies—and hence quite different mobilities. For $N = 10$–23, a single isomer, with a prolate topology is present for each cluster species At a nuclearity of $N = 24$, there is a second series of isomers with greater mobilities, which have been attributed to oblate clusters. Beyond $N = 35$, only oblate isomers are present.

Jarrold also performed annealing studies on the silicon clusters, by increasing the injection energy, so that collisions between the cluster ions and the helium carrier gas impart greater internal energy to the cluster, which in turn can lead to cluster reconstruction. For Si_{32}^+, for example, at the higher injection energy, the slow moving component (the prolate isomer) disappears, as it anneals into the more spherical oblate isomer. In the region $N = 25$–29, both isomer types are still observed after annealing, but from $N = 30$–35, annealing results in the complete loss of the prolate isomer— indicating that the prolate isomers are thermodynamically unstable with respect to the oblate isomers in this size regime.

CHAPTER 7

7.1 In ion-mobility experiments, the drift time is proportional to the effective cross sectional area (CSA) of the cluster as it rotates, since 'larger' molecules take part in more collisions with the buffer gas. Assuming that there is no preferred alignment of the molecules, the CSA is determined by the length of the longest inertial axis of the cluster. The '8×3×3' isomer of $(NaCl)_{35}Cl^-$ is longer and thinner (more prolate) than the '6×4×3' isomer and, hence, carves out a larger sphere as it rotates, giving it a lower mobility and higher drift time, as shown in Section 7.3.1. The '6×6×2' isomer is oblate, but as the longest axis is the same as for the '6×4×3' isomer, there should not be much difference in mobility, though it should have a shorter drift time than the '8×3×3' isomer and a longer drift time than the more pseudo-spherical '5×5×3' isomer.

7.2 Mass spectroscopy gives information as to the stability of ionic clusteres as a function of size. The observed magic numbers can be interpreted in terms of low energy structures, often having complete cuboidal geometries in the case of alkali halide and MgO clusters. Ion mobility measurements (see Q7.1) can also be used to distinguish between possible isomers with distinct geometries. (See Section 7.3.1 for more details.)

Index